Advances in the Immunopathogenesis of Multiple Sclerosis

Springer
Milano
Berlin
Heidelberg
New York
Barcelona
Hong Kong
London
Paris
Singapore
Tokyo

Advances in the Immunopathogenesis of Multiple Sclerosis

Editors:
D. Gambi
P.A. Muraro
A. Lugaresi

Guest Editor:
U. Ecari

 Springer

D. GAMBI
P.A. MURARO
A. LUGARESI
Department of Oncology
and Neuroscience
Medical School
University G. D'Annunzio
Chieti, Italy

The Editors and Authors wish to thank FARMADES-SCHERING GROUP (Italy)
for their support and help in the realization and promotion of this volume

© Springer-Verlag Italia, Milan 1999

ISBN 978-88-470-0067-4   ISBN 978-88-470-2269-0 (eBook)
DOI 10.1007/978-88-470-2269-0

Library of Congress Cataloging-in-Publication Data: Applied for

Cover design: Simona Colombo, Milan
Typesetting: Graphostudio, Milan

SPIN: 10737528

This volume is dedicated to the memory of Professors Giorgio Macchi and Luigi Amaducci. Through their studies in neuroscience, these professors set the foundation for the modern scientific approach to multiple sclerosis in Italy. Their work is kept alive today by those neurologists who had the opportunity to learn from them.

Table of Contents

Contributors

M. Arese
Department of Genetics, Biology and
Biochemistry, and Institute for Cancer
Research and Treatment, University of
Turin, Italy

L. Battistini
Laboratory of Neuroimmunology, IRCCS
S. Lucia, Rome, Italy

A. Bergami
Experimental Neuroimmunotherapy Unit,
Department of Neurology, DIBIT,
Scientific Institute San Raffaele, Milan,
Italy

B. Bielekova
Neuroimmunology Branch, National
Institute of Neurological Disorders and
Stroke, NIH, Bethesda, MD, USA

R. Bomprezzi
First Chair of Neurology, Department of
Neurosciences, University La Sapienza,
Rome, Italy

G. Borsellino
Laboratory of Neuroimmunology, IRCCS
S. Lucia, Rome, Italy

H. Brok
Department of Immunobiology,
Biomedical Primate Research Centre,
Rijswijk, The Netherlands

F. Bussolino
Department of Genetics, Biology and
Biochemistry, and Institute for Cancer
Research and Treatment, University of
Turin, Italy

C. Buttinelli
First Chair of Neurology, Department of
Neurosciences, University La Sapienza,
Rome, Italy

L. Callea
Department of Neurology, Civil Hospitals,
Brescia, Italy

S. Cannoni
First Chair of Neurology, Department of
Neurosciences, University La Sapienza,
Rome, Italy

E. Capello
Department of Neurological Sciences,
University of Genoa, Italy

G. Desina
Experimental Neuroimmunotherapy Unit,
Department of Neurology, DIBIT, Scientific
Institute San Raffaele, Milan, Italy and
Department of Neurology, Casa Sollievo
della Sofferenza Scientific Institute, San
Giovanni Rotondo (FG), Italy

C. Ferrandi
Department of Genetics, Biology and
Biochemistry, and Institute for Cancer
Research and Treatment University of
Turin, Italy

M. Filippi
Neuroimaging Research Unit, Department
of Neuroscience, Scientific Institute San
Raffaele, Milan, Italy

R. Furlan
Experimental Neuroimmunotherapy Unit,
Department of Neurology, DIBIT, Scientific
Institute San Raffaele, Milan, Italy

D. Gambi
Department of Oncology and
Neuroscience, Medical School, University
G. D'Annunzio, Chieti, Italy

M. Gironi
Experimental Neuroimmunotherapy
Unit, Department of Neurology, DIBIT,
Scientific Institute, San Raffaele, Milan,
Italy

D. Giunti
Department of Neurological Sciences,
University of Genoa, Italy

B. Gran
Cellular Immunology Section,
Neuroimmunology Branch, National
Institutes of Neurological Disorders and
Stroke, NIH, Bethesda, MD, USA

B. Hemmer
Neuroimmunology Section, Department of
Neurology, Philipps-University of
Marburg, Germany and Cellular
Immunology Section, Neuroimmunology
Branch, National Institutes of Neurological
Disorders and Stroke, NIH, Bethesda, MD,
USA

R. Houghten
Multiple Peptide Systems and Torrey Pines
Institute for Molecular Studies, San Diego,
CA, USA

G. Iannucci
Neuroimaging Research Unit, Department
of Neuroscience, Scientific Institute San
Raffaele, Milan, Italy

A. Lugaresi
Department of Oncology and
Neuroscience Medical School, University
G. D'Annunzio, Chieti, Italy

G.L. Mancardi
Department of Neurological Sciences,
University of Genoa, Italy

R. Martin
Cellular Immunology Section,
Neuroimmunology Branch, National
Institutes of Neurological Disorders and
Stroke, NIH, Bethesda, MD, USA

G. Martino
Experimental Neuroimmunotherapy Unit,
Department of Neurology, DIBIT, Scientific
Institute San Raffaele, Milan, Italy

L. Massacesi
Department of Neurological and
Psychiatric Sciences, University of
Florence, Italy

B. Mazzanti
Department of Neurological and
Psychiatric Sciences, University of
Florence, Italy

H.F. McFarland
Cellular Immunology Section,
Neuroimmunology Branch, National
Institutes of Neurological Disorders and
Stroke, NIH, Bethesda, MD, USA

I. Medana
Neuroimmunology, Max-Planck Institute
of Neurobiology, Martinsried, Germany

T. Misgeld
Neuroimmunology, Max-Planck Institute
of Neurobiology, Martinsried, Germany

C. Montesperelli
First Chair of Neurology, Department of
Neurosciences, University La Sapienza,
Rome, Italy

P. A. Muraro
Department of Oncology and
Neuroscience, Medical School, University
G. D'Annunzio, Chieti, Italy and
Neuroimmunology Branch, National
Institutes of Neurological Disorders and
Stroke, NIH, Bethesda, MD, USA

H. Neumann
Neuroimmunology, Max-Planck Institute
of Neurobiology, Martinsried, Germany

A. Perna
First Chair of Neurology, Department of
Neurosciences, University La Sapienza,
Rome, Italy

C. Pinilla
Torrey Pines Institute for Molecular
Studies, San Diego, CA, USA

P.L. Poliani
Experimental Neuroimmunotherapy Unit,
Department of Neurology, DIBIT, Scientific
Institute San Raffaele, Milan, Italy

C. Pozzilli
First Chair of Neurology, Department of
Neurosciences, University La Sapienza,
Rome, Italy

G. Ristori
First Chair of Neurology, Department of
Neurosciences, Università La Sapienza,
Rome, Italy

L. Roccatagliata
Department of Neurological Sciences,
University of Genoa, Italy

M. Rovaris
Neuroimaging Research Unit, Department
of Neuroscience, Scientific Institute San
Raffaele, Milan, Italy

M. Salvetti
First Chair of Neurology, Department of
Neurosciences, University La Sapienza,
Rome, Italy

B. t'Hart
Department of Immunobiology,
Biomedical Primate Research Centre,
Rijswijk, The Netherlands

E. Traggiai
Department of Neurological and
Psychiatric Sciences, University of
Florence, Italy

A. Uccelli
Department of Neurological Sciences,
University of Genoa, Italy

M. Vergelli
Department of Neurological and
Psychiatric Sciences, University of
Florence, Italy

Introduction

P.A. MURARO, A. LUGARESI, D. GAMBI

Many of the pathological aspects of multiple sclerosis (MS) lesions have been known for over a century. It is only recently, however, that different patterns of demyelination have been linked to distinct pathways of immune-mediated tissue destruction. In particular, the inter-individual heterogeneity of MS lesions has suggested that different mechanisms may act in different patients, accounting for the variability observed in clinical course, immunological findings in peripheral blood and cerebrospinal fluid (CSF), and response to immunomodulatory treatments.

To provide an overview of the basic mechanisms possibly involved in MS lesion initiation and development, an international meeting was organized in the context of the annual Congress of the Italian Neuroimmunology Association (AINI), held at the University of Chieti, in Chieti Italy on 29 October 1998. The high standard of presentations prompted us to report them in extended form, to highlight recent progress in the understanding of basic mechanisms sustaining MS immuno-pathogenesis.

A central role in the possible mechanisms leading to myelin destruction has been attributed to T lymphocytes reactive to myelin antigens. Studies on the myelin antigen-specific T cell repertoire have contributed significant advances to our knowledge of autoimmunity (Chapters 1, 2). Which mechanisms might be responsible for the recruitment of myelin-specific T cells in MS remains the main unsolved issue. In this context, it has been proposed that cross-reactivity with foreign antigens may lead to the activation of autoantigen-specific T lymphocytes, which are a normal component of the T cell repertoire. This concept, known as the molecular mimicry hypothesis, is currently the focus of intense research efforts (Chapters 3, 4).

In addition to the antigenic stimulus, an essential role in T cell activation is played by major histocompatibility complex (MHC) molecules, which present peptide antigens to T cells. The modulation of MHC expression by neuronal cell-derived mediators has suggested the possible involvement of resident central nervous system (CNS) cells in antigen presentation to myelin-reactive cells (Chapter 5).

Alterations of the blood-brain barrier (BBB) may also be required for the recruitment of autoreactive lymphoid cells into the CNS. Contributing to the disruption of the barrier function of the brain vascular endothelium are soluble inflammation mediators, which may have the ability to impair the function of the endothelial cell monolayer (Chapter 6). The importance of BBB dysfunction in MS, first established by the immunological abnormalities found in the CSF, has been confirmed by magnetic resonance imaging studies (Chapter 7). Studies on experimental allergic encephalomyelitis (EAE) have demonstrated the importance of cytokines in the peripheral activation of lymphocytes required to obtain perivascular infiltration into the CNS. In addition, proinflammatory cytokines may also be involved in the effector phase of inflammation and demyelination (Chapter 8). EAE studies in outbred animals aim to understand the mechanisms of inflammation and demyelination in a model potentially resembling more closely the human situation in MS (Chapter 9).

We hope this volume will provide helpful insight into the pathogenetic mechanisms of demyelination in MS. We would like to thank all the authors who have made this publication possible, and wish them success in their future efforts.

Chapter 1

The affinity spectrum of myelin basic protein-reactive T cells

B. Mazzanti, E. Traggiai, B. Hemmer, R. Martin, L. Massacesi, M. Vergelli

Introduction

T cell reactivity to myelin basic protein in multiple sclerosis

The breakdown of the immunological tolerance to myelin components is probably crucial in the pathogenesis of multiple sclerosis (MS). Even if the target autoantigen is yet unknown, myelin basic protein (MBP) has been studied in greatest detail [1, 2].

Although it was not expected at the beginning, MBP-specific, potentially autoreactive T cells are present in the peripheral blood (PB) of MS patients as well as in normal individuals [3, 4]. Similar autoreactive T cells were found in the peripheral lymphoid organs of experimental animals; upon in vitro activation these cells mediated autoimmune central nervous system (CNS) damage [5, 6]. These findings indicate that autoantigen-reactive T cells escape the negative selection process in the thymus and belong to the "normal" T cell repertoire.

Although several studies suggest the importance of MBP-reactive T cells in the pathogenesis of MS [7, 8], it is still controversial whether MBP-specific T cells derived from MS patients differ with respect to frequency and/or functional properties in comparison to MBP-specific T cells derived from healthy individuals [9].

It has been recently reported that T cells specific for MBP, even if derived from the same subject, are highly heterogeneous in terms of functional phenotype including the antigen requirement to become activated [10]. The differences in antigenic doses necessary for activation most likely relate to differences in the affinity of the T cell receptor (TCR) for the specific major histocompatibility complex (MHC)-peptide complex. This implies that T cells with differential TCR affinities constitute the repertoire of MBP-reactive T cells. However, the relations between functional characteristics and affinity of the T cell response to

MBP in MS, as well as for other autoantigens in autoimmune diseases, have not been addressed.

Degeneracy in antigen recognition by MBP-specific T cells

In the last few years it has become evident that T cell recognition is highly flexible [11, 12] and that a high level of cross-reactivity is an essential feature of the T cell receptor [13]. In this context, recent studies from our group have shown a high degree of degeneracy in antigen recognition by MBP-reactive T cells [14-16]. The identification of single amino acid substitutions of MBP peptides yielding increased T cell responsiveness (superagonist substitutions) suggested that the number of productive ligands for a single autoreactive T cell clone is much higher than previously expected [14]. In addition, studies with modified peptides carrying multiple amino acid substitutions indicated that none of the residues of the peptide is strictly necessary for T cell recognition: the negative effects of any substitution can be compensated by modifications inducing superagonist activity at other positions [14, 15]. Based on these results, cross-reactive ligands which do not share any amino acid with the "wild type" peptide antigen were synthesized [15].

By extending these findings to their limit, the use of combinatorial peptide libraries (consisting of mixtures containing up to 10^{14} different peptides) definitely demonstrated the extent of degeneration in antigen recognition by MBP-reactive T cell clones [16]. This approach allowed the identification of cross-reactive peptides inducing T cell responses at lower antigen concentration compared to the native MBP peptide. For any T cell clone, an affinity hierarchy can be envisioned with a small number of peptide antigens which function as optimal ligands and a wide range of suboptimal ligands recognized with lower affinity. In this context it may be hypothesized that the "real" specificity of MBP-reactive T cells is different from the antigen used for the selection and the expansion of T cells in culture. The MBP peptide could just happen to be a low-affinity cross-reactive ligand.

Decrypting the repertoire of antigen-specific T cells by using different antigen concentrations in culture

Considering that the use of low and high antigen concentrations resulted in the selection of T cells with different phenotypes in various in vivo and in vitro systems [17-19], we asked whether changes in the antigen concentration (which in turn result in different numbers of MHC/peptide complexes on the surface of antigen presenting cells) influence the expansion of the antigen-specific T cell repertoire in vitro.

MBP

Short-term MBP-specific T cell lines were generated from the peripheral blood of 15 subjects (8 MS patients and 7 normal donors) by using a modified split-well technique [20]. The primary cultures were started by seeding 2 x 10^5 peripheral blood mononuclear cells with different concentrations of MBP (0.1-50 µg/ml). Seven days later 10 U/ml IL-2 was added to each well. At the fourteenth day, the content of each well was split into 2 new wells. One well was stimulated with the appropriate antigen concentration, the other with medium alone. Three days later, 1 µCi ^{3}H-thymidine was added to each well and after 6 additional hours ^{3}H-thymidine incorporation was measured in a scintillation counter. Wells showing a stimulation index >2 were considered positive. In spite of the variability among individuals, the frequency of MBP-reactive wells was strictly dependent upon the antigen concentration used in the primary cultures for each subject. No significant difference in the estimated frequency of MBP-reactive T cells was observed between MS patients and healthy donors at any antigen concentration. At low antigen concentrations (<1 µg/ml) no MBP-reactive T cell lines could be isolated from either MS patients or healthy donors. The mean percentage of positive wells in the cultures treated with 50, 10 and 1 µg/ml MBP were 63.6%, 22.2% and 5.3%, respectively. These results indicate that the precursor frequency of antigen-specific T cells is not a fixed number but strongly depends upon the antigen dose. A high frequency of MBP-reactive T cells can be detected in the peripheral blood of both MS patients and healthy individuals by using high antigen concentrations in vitro.

The fine specificity of T cell lines derived from 4 individuals (3 MS patients and 1 healthy donor) at different MBP concentrations was evaluated by using a panel of seventeen 20-mer synthetic peptides spanning the entire sequence of human MBP. For each individual, various immunodominant peptides were identified. However, within the same subject, the use of different MBP concentrations for generating T cell lines selected T cells with different specificities. T cell lines generated at higher antigen concentrations recognized a wider panel of peptide epitopes compared to T cells raised at low antigen doses, indicating that the response to MBP is progressively less focused with increasing antigen doses.

Finally, a number of T cell lines raised with different antigen concentrations from two individuals were tested for their proliferative response to increasing concentrations of MBP. The antigen dose requirements of MBP-reactive T cells was related to the antigen amount used in the primary cultures. T cell lines generated at low antigen concentration (1 µg/ml) reached the half-maximal response (EC_{50}) at lower antigen dose than T cell lines generated at high MBP concentration (50 µg/ml). These results indicate that the use of different antigen concentrations in the primary culture allows the selection of T cells with different antigen requirements.

Flu-HA

Based on these findings it can be speculated that calculation of the frequency of specific T cells at different antigen doses is a useful tool in estimating the overall affinity repertoire of a given T cell response. In this context, it is relevant that similar results were obtained when a foreign antigen such as influenza virus hemoagglutinin peptide 306-318 (Flu-HA 306-318) was used for the generation of T cell lines.

Short-term antigen-specific T cell lines were established by the split-well technique, by culturing peripheral blood mononuclear cells with different Flu-HA 306-318 concentrations (range 10^{-5}-10 µg/ml). A valuable percentage of Flu-HA 306-318-reactive T cell lines (10%) was isolated using concentrations as low as 10^{-5} µg/ml. A dose-dependent increase in the percentage of Flu-HA 306-318-reactive T cells was observed in the concentration range 10^{-4}-10^{-2} µg/ml. A concentration of 10^{-3} µg/ml yielded over 70% positive cultures. Using higher Flu-HA 306-318 concentrations (10^{-2}-10 µg/ml), all the cultures were positive.

In addition, bulk cultures generated in vitro with different antigen doses were tested for proliferative response to increasing concentrations of Flu-HA 306-318. The antigen dose requirement of Flu-HA 306-318-reactive T cells was dependent upon the antigen amount used in the primary cultures. This behavior determines that the EC_{50} is reached at lower concentrations for the culture grown at low antigen doses.

These data show in a different experimental system that the estimated frequency of antigen-reactive T cells is dependent upon the antigen concentration. Moreover, an inverse relationship between antigen dose in the primary culture and antigen requirement (affinity) of T cells was found. It is interesting that the maximal reactivity was observed at antigen concentrations three orders of magnitude lower than that necessary for the MBP response.

Taken together, these findings indicate that (1) the frequency of antigen-specific T cells is dependent upon the antigen concentration used for the generation of antigen-specific T cell lines; and (2) the antigen concentration used in vitro in the primary culture selects for T cell populations with different antigen dose requirements. In this experimental framework, lower concentrations of the antigen yielded a smaller number of T cell lines recognizing the specific antigen with higher affinity, whereas high antigen concentrations allowed the in vitro expansion of a higher numbers of T cells recognizing their specific ligand with lower affinity.

Looking at these results in the context of degenerate T cell recognition and its implications for cross-reactivity, it is conceivable that the in vitro generation of MBP-specific T cells by using high concentrations of autoantigen is actually reading out the potential for low-affinity cross-reactivity rather than the "real" antigen specificity of the responding cells. The higher the antigen dose used, the lower the affinity for MBP and the further from MBP is the "real specificity" of the selected T cells. Conversely, at low antigen doses, one would select T cells whose optimal peptide sequence is closer to MBP.

Molecular mimicry, natural autoreactivity and autoimmunity: a unifying hypothesis

The findings herein reported indicate that a high number of low-affinity autoreactive T cells is part of the "normal" T cell repertoire. Due to the large extent of degeneracy and cross-reactivity in T cell recognition, low-affinity, self-reactive T cells can be commonly activated and expanded during the course of protective immune reactions to other (foreign) agents. This frequently occurring cross-recognition is unlikely to be "dangerous" since low-affinity antigen recognition may prevent activation of these autoreactive T cells even if they encounter the cross-reactive self antigen within the target organ (i.e. the central nervous system). Additional events may be required for "normal" autoreactivity to become frank autoimmunity. In the presence of a concomitant inflammatory process in the CNS leading to increased expression of MHC and costimulatory molecules, the low affinity can be compensated by the elevated number of available MHC-peptide complexes on the surface of antigen-presenting cells and by costimulatory factors raising the overall avidity beyond the threshold required for triggering an effector T cell response. In this context low affinity autoreactive T cells may acquire pathogenic properties.

In summary, this model may explain why it has been so difficult to identify differences in the frequency of self-antigen reactive T cells in patients with autoimmune diseases in comparison to normal donors. Autoreactivity can in fact be considered an intrinsic feature of the normal immune system. The requirement for additional factors occurring concomitantly to the cross-activation of self-reactive T cells may also explain why autoimmunity is relatively rare.

References

1. Martin R, McFarland HF, McFarlin DE (1992) Immunological aspects of demyelinating diseases. Annu Rev Immunol 10:153-187
2. Hohlfeld R (1997) Biotechnological agents for the immunotherapy of multiple sclerosis. Principles, problems and perspectives. Brain 120:859-916
3. Burns J, Rosenzweig A, Zweiman B, Lisak RP (1983) Isolation of myelin basic protein-reactive T cell lines from normal human blood. Cell Immunol 81:435-440
4. Pette M, Fujita K, Kitze B, Whitaker JN, Albert E, Kappos L, Wekerle H (1990) Myelin basic protein-specific T lymphocyte lines from MS patients and healthy individuals. Neurology 40:1770-1776
5. Schlüsener H, Wekerle H (1985) Autoaggressive T lymphocyte lines recognize the encephalitogenic region of myelin basic protein; in vitro selection from unprimed rat T lymphocyte populations. J Immunol 135:3128-3133
6. Genain CP, Lee-Parritz D, Nguyen M-H, Massacesi L, Joshi N, Ferrante R, Hoffman K, Moseley M, Letvin NL, Hauser SL (1994) In healthy primates, circulating autoreactive T cells mediate autoimmune disease. J Clin Invest 94:1339-1345
7. Steinman L, Waisman A, Altman D (1995) Major T cells responses in multiple sclerosis. Mol Med/Today 1:79-83
8. Wucherpfennig KW, Weine HL, Hafler DA (1991) T cells recognition of myelin basic protein. Immunol Today 12:277-282
9. Martin R, McFarland HF (1995) Immunological aspects of experimental allergic encephalomyelitis and multiple sclerosis. Crit Rev Clin Lab Sci 32:121-182
10. Hemmer B, Vergelli M, Tranquill L, Conlon P, Ling N, McFarland HF, Martin R (1997) Human T cell response to myelin basic protein peptide (83-99): extensive heterogeneity in antigen recognition, function and phenotype. Neurology 49:1116-1126
11. Kersh G, Allen PM (1996) Essential flexibility in the T cell recognition of antigen. Nature 380:495-498
12. Hemmer B, Vergelli M, Pinilla C, Houghten RA, Martin R (1998) Probing degeneracy in T cell recognition using peptide combinatorial libraries - Importance for T cell survival and autoimmunity. Immunol Today 19:163-168

13. Mason D (1998) A very high level of crossreactivity is an essential feature of the T cell receptor. Immunol Today 19:395-404
14. Vergelli M, Hemmer B, Kalbus M, Vogt AB, Ling N, Conlon P, Coligan JE, McFarland HF, Martin R (1997) Modifications of peptide ligands enhancing T cell responsiveness suggest a broad spectrum of stimulatory ligands for autoreactive T cells. J Immunol 58:3746-3752
15. Hemmer B, Vergelli M, Gran B, Ling N, Conlon P, Pinilla C, Houghten R, McFarland HF, Martin R (1998) Predictable T cell receptor antigen recognition based on peptide scans leads to the identification of agonist ligands with no sequence homology. J Immunol 160:3631-3636
16. Hemmer B, Fleckenstein BT, Vergelli M, Jung W, McFarland HF, Martin R, Wiesmuller KH (1997) Identification of high potency microbial and self ligands for a human autoreactive class II-restricted T cell clone. J Exp Med 185:1651-1659
17. Constant S, Pfeiffer C, Woodard A, Pasqualini T, Bottomly K (1995) Extent of T cell receptor ligation can determine the functional differentiation of naive CD4+ T cells. J Exp Med 182:1591-1596
18. Kersh GJ, Donermeyer DL, Frederick KE, White JM, Hsu BL Allen PM (1998) TCR transgenic mice in which usage of transgenic alpha- and beta-chains is highly dependent on the level of selecting ligand. J Immunol 161:585-593
19. Rogers PR, Huston G, Swain SL (1998) High antigen density and IL-2 are required for generation of CD4 effectors secreting Th1 rather than Th0 cytokines. J Immunol 161:3844-3852
20. Vergelli M, Hemmer B, Utz U, Vogt A, Kalbus M, Tranquill L, Conlon P, Ling N, Steinmann L, McFarland HF, Martin R (1996) Differential activation of human autoreactive T cell clones by altered peptide ligands derived from myelin basic protein peptide (87-99). Eur J Immunol 26:2624-2634

The naïve and memory MBP-reactive CD4+ T cell repertoire: implications for the autoimmune concept in multiple sclerosis

P.A. Muraro, B. Bielekova

T cell responses to myelin basic protein in MS

T cells reacting to self-antigens may be involved in the pathogenesis of human autoimmune diseases, such as multiple sclerosis (MS). The pathogenesis of MS is currently ascribed, at least in part, to a T cell-mediated process targeting myelin components [1]. Among myelin proteins, myelin basic protein (MBP) is a major candidate autoantigen in MS. The evidence supporting this claim has been examined in several excellent reviews [1-4], and includes the following well established observations: (1) degradation products of MBP are present in macrophages surrounding MS lesions; (2) MBP content in plaques is reduced; (3) the protein is relatively abundant (~30%) in central nervous system (CNS) myelin and (4) the immunodominant T cell epitope clusters of MBP [N-terminal (1-11, 1-19), central (87-106, 83-99, 87-99, 84-102), central (111-129, 115-125), C-terminal (143-168, 131-159)] are predominantly recognized in the context of MS-associated HLA-DR molecules [5-10].

As additional support to a possible role of MBP responses in MS, MBP epitopes corresponding to human immunodominant regions are encephalitogenic in animal models (experimental allergic encephalomyelitis, EAE). Epitopes which can induce EAE vary among different animal species and strains, depending on the immunogenetic background [1]. Consistent with this observation in EAE, we have recently demonstrated in humans a close relationship between the immunodominance of MBP epitopes and the human leukocyte antigen (HLA) background of the population examined [10]. Although myelin-specific cells can mediate EAE under the appropriate conditions, these cells are found as part of the normal human T cell repertoire in both MS patients and healthy subjects at comparable frequencies [5, 7, 10, 11], raising questions about their actual involvement in the disease process.

Despite this discrepancy, it was only recently that studies on the frequency of self-reactive T lymphocytes considered whether these cells had been antigen-primed in the body, i.e. whether they derived from a "memory" or a "naïve" pool. Answering this question would afford a better understanding of the possible involvement of myelin-specific cells in MS. Functional assays have demonstrated the presence of MBP-specific responses independent from CD28/B7-mediated costimulation, which were attributed to a memory reactivity [12, 13]. These approaches, however, did not discriminate between a long-term recall response and the presence of recently activated effector cells responding to MBP. For this reason, we have investigated the origin of MBP reactive cells in the naïve vs. memory compartments as identified by their distinctive phenotypes.

Autoreactivity and immunological memory

Currently available data suggest that autoreactive cells normally found in the immune system, such as MBP-specific T cells, may become activated and target self components upon encounter of cross-reactive antigens in peripheral lymphoid organs [14, 15]. Accordingly, in patients with autoimmune disease, autoreactive cells are expected to be initially recruited from the naïve cell pool and become antigen-primed upon activation and repeated antigenic contact in the target organ. On these grounds, autoreactive cells are expected to retain an immunological memory, which can be defined as the ability to generate a more effective immune response after a second encounter with the antigen [16].

The ability of memory T cells to respond more vigorously than virgin T cells to the same activation signal is a fundamental property of acquired immunity [17, 18]. In fact, primed cells do respond better than naïve cells to anti-CD2 or -CD3 monoclonal antibodies (mAbs), while naïve cells respond equally well or better to mitogens. The greater efficiency in the response to antigen of memory cells can be achieved in the absence of higher affinity receptors, by means of an increased expression of adhesion molecules that influence cell-to-cell interactions. This assumption is supported by the differential expression of a number of cell surface molecules which are expressed on virgin and memory T cells [19, 20]. Enhanced expression of these markers might facilitate the interaction of memory T cells with antigen presenting cells (APC) or alter the type(s) of APC with which memory T cells are capable of interacting [21-23]. Moreover, differential expression of certain surface cell markers may alter the recirculation pathways of virgin and memory

T cells, facilitating the reencounter of memory cells with their antigens in the same body compartment where the initial priming occurred [19, 20, 24, 25].

Upon antigenic contact, human T cells undergo several phenotypic changes, including the reciprocal downregulation of the high-molecular weight CD45RA isoform and upregulation of the low-molecular weight form CD45RO. Although the switch of CD45R isotypes can be reversible, at least in rats [26, 27], and may not occur reliably in CD8$^+$ cells [28], these molecules have been considered for several years the best available markers and have been widely used for the identification and selection of naïve (CD45RA$^+$/RO$^-$) and memory (CD45RA$^-$/RO$^+$) T cells [29, 30]. Therefore, we applied this concept to investigate the origin of MBP responses from naïve and memory CD4$^+$ T cell subsets.

Naïve and memory MBP-specific T cell repertoires in MS

In order to better understand the origin, the specificities and the functional characteristics of CD4$^+$ MBP-specific T cells derived from the peripheral blood of MS patients and controls, we asked the following questions: (a) Which is the frequency of MBP-specific T cells originating from the putative naïve (CD45RA$^+$/RO$^-$) and memory (CD45RA$^-$/RO$^+$) CD4$^+$ subsets? (b) Do autoreactive cells selected from the naïve and memory subsets respond to different antigenic epitopes? (c) Are the MBP-reactive naïve and memory T cell repertoires different with respect to their antigen affinity?

Partitioning of the T cell response to MBP in the naïve and memory compartments

To estimate the frequency of autoantigen-specific T cells in the naïve and memory populations, we isolated CD4$^+$ CD45RA$^+$/RO$^-$ and CD45RA$^-$/RO$^+$ T cells by negative selection from 7 MS patients and 4 healthy donors, and generated MBP-specific T cell clones (TCC) from the highly purified(≥95%) subsets. Cells from both subsets proliferated comparably well in response to potent stimuli such as mitogens or superantigens. On the contrary, it was predominantly the naïve T cell population which responded to MBP in the majority of both MS patients and healthy controls (Fig. 1). Once excluded a clonally expanded cell populations, the estimated precursor frequency of MBP-reactive T cells in the naive subset was significantly higher than in the memory subset both of MS patients, ($p = 0.040$, t test) and of healthy donors ($p = 0.024$).

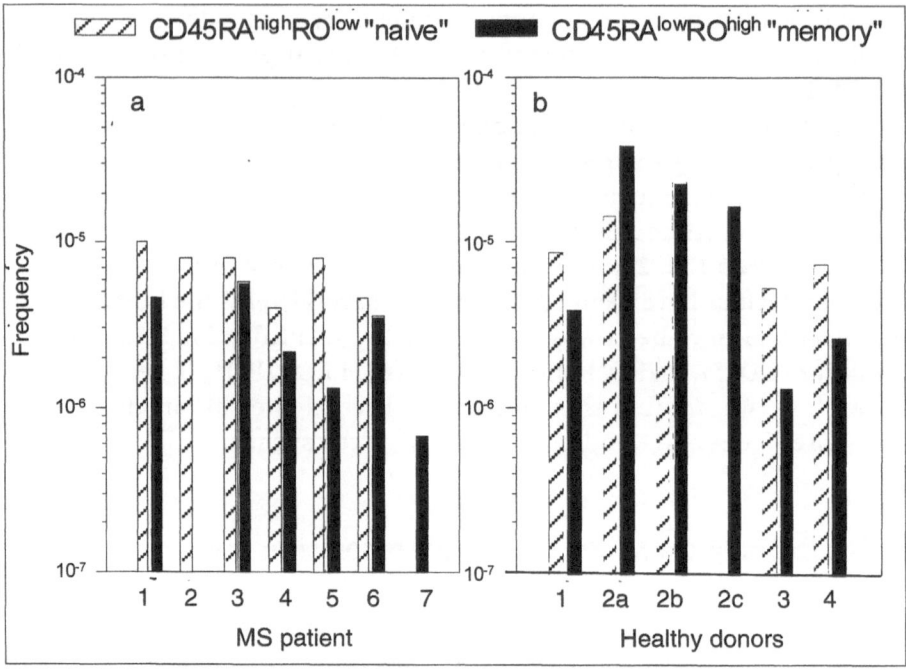

Fig. 1a, b. *Estimated precursor frequency of MBP-reactive T cells in the naïve and memory CD4⁺ T cell subsets.* The precursor frequency of MBP-specific T cell lines originating from CD45RA⁺/RO⁻ naïve (*dashed bars*) and CD45RA⁻/RO⁺ memory (*black bars*) T cell subsets was calculated by limiting dilution analysis. In the majority of MS patients (**a**) and healthy donors (**b**), MBP responses originated prevalently from the CD45RA⁺/RO- naïve subset. As an exception, a clear prevalence of MBP responses from the CD45RA⁻/RO⁺ memory pool was found in one healthy donor (*No. 2*). In this subject, a prevalent reactivity from the memory cell subset was found when cells were stimulated with MBP (*2a*) or with the DRB1 0401-restricted immunodominant peptide MBP (111-129) (*2b*). This was observed even to a larger extent in a second experiment, in which stimulation with MBP (111-129) yielded a memory-derived response only (*2c*). Since these responses were sustained by a single, unusually expanded TCC [10], they were not included in the statistical analysis of the overall data

Recognition of different antigenic epitopes by MBP-specific TCC from the naïve and memory subsets

We next assessed the antigen fine specificities of TCC originating from the naïve and memory subsets in proliferation assays, using a panel of overlapping MBP peptides. T cell lines (TCL) specific for the immunodominant epitopes MBP(61-79, 81-99 and 141-159) predominantly emerged from the naïve CD4⁺ subset. Notably, 13 of 14 TCC (among which, 7 of 8 were MS patients) responding to MBP(81-99), the immunodominant region of MBP in DR2+ individuals, derived from the naïve subset. The MBP epitopes recognized by TCC from either subset excluded each other almost completely in every donor, demonstrating a clear skewing in the specificities of TCC originated from the two different cell compartments.

Higher functional antigenic affinity of naïve subset-derived MBP-specific TCC

To evaluate the antigenic affinity of naïve vs. memory subset-originating MBP-specific T cells, we measured the antigen concentration required to induce half-maximal responses to MBP peptides (EC_{50}) of a representative panel of TCC. Naïve-derived TCC had significantly lower EC_{50} than did memory-derived TCC ($p < 0.001$, Mann-Whitney rank sum test). This did not depend on a greater proliferation potential to nonspecific stimuli nor on a higher expression of adhesion molecules or ligands for costimulation, indicating selection from the naïve repertoire of TCC with a higher functional avidity for MBP.

Naïve responses to MBP in MS or immune memory within the CD45RA+ subset?

By dissecting the response to MBP generated from highly purified CD45RA+ and CD45RO+ human CD4+ T cell subsets, we found a predominant reactivity from cells originally showing a naïve CD45RA+ phenotype, both in MS patients and controls. Moreover, T cells specific for immunodominant MBP epitopes originated almost exclusively in the CD45RA+ subset. Finally, CD45RA+ subset-derived MBP-specific T cells had a higher functional affinity for the antigen than those raised from the CD45RO+ subset. The conclusions which can be drawn from these observations will necessarily depend on how reliably one can attribute a naïve or memory *origin* to cells showing a conventional naïve or memory *phenotype*. A primed immune response to MBP is supported in MS by the presence of in vivo activated [31, 32] and costimulation-independent [12, 13] MBP-specific T cells. These observations may argue against the possibility of a purely naïve response to MBP in MS. As previously mentioned, it has been shown, particularly in a rat system, that CD4+ cells expressing the low-molecular weight CD45R isoform (corresponding to the human CD45RO) can revert to a naïve phenotype after withdrawal from antigenic contact [26, 27]. Observations in human patients treated with radiation therapy also indicated that CD45R high-molecular

weight-expressing T cells might represent both virgin and primed populations which have reverted to a naïve phenotype [33]. Thus, it cannot be
excluded that after initial expansion, memory cells reacting to immunodominant MBP epitopes might revert to a CD45RA$^+$ phenotype upon
antigen deprivation in the peripheral blood, thus explaining the
increased frequency we observed in that subset. As the only detected
clue to a recall response, these cells, otherwise so far phenotypically
indistinguishable from naïve cells, show an increased functional affinity
for the antigen.

Ongoing efforts in our laboratory are aimed to investigate further
and in a larger scale the functional characteristics of the prevalent,
immunodominant MBP responses from the CD45RA$^+$/RO$^-$ CD4$^+$ T cell
compartment, which point to a central role of this subset as a source of
autoreactivity in MS.

Summary

Multiple sclerosis (MS) is considered to be an autoimmune disorder
mediated by CD4$^+$ T cells reactive to myelin antigens, including myelin
basic protein (MBP). In order to address the naïve vs. memory origin of
circulating myelin-reactive T cells, we have generated MBP-specific T
cell clones (TCC) from the putative naïve (CD45RA$^+$RO$^-$) and memory
(CD45RO$^+$RA$^-$) CD4$^+$ T cell subsets, purified from the peripheral blood
of MS patients and healthy donors. Our results demonstrate that: (a) the
T cell response to MBP in MS patients predominantly derives from the
naïve CD4$^+$ T cell subset; (b) in each individual, the recognition of single
MBP epitopes is consistently skewed to either the naïve or memory subset; (c) the specific reactivity to the immunodominant MBP epitopes, in
particular MBP(81-99) and MBP(141-159), originates predominantly
from the naïve subset; and (d) the affinity for the antigen of naïve subset-derived TCC is higher than that of memory-derived TCC.

Significant systematic differences characterize MBP-specific TCC
originating from the two T cell subsets. Taken together, these observations indicate that immunodominant epitope-specific cells from the
CD45RA$^+$ repertoire may have a greater potential to contribute to the
"effector" cell pool than subdominant epitope-specific cells from the
CD45RO$^+$ subset. The possibility of a presence of immunological memory in the phenotypically naïve CD4$^+$ CD45RA$^+$ MBP-specific T cell
repertoire may have important implications for the understanding of
autoimmunity.

Acknowledgement
We thank Dr. Martin Pette for his early contribution, essential to undertaking this study. We also would like to thank our mentors Drs. Roland Martin and Henry F. McFarland for their invaluable advice and support.

References

1. Martin R, McFarland HF, McFarlin DE (1992) Immunological aspects of demyelinating diseases. Annu Rev Immunol 10:153-187
2. Hafler DA, Weiner HL (1995) Immunologic mechanisms and therapy in multiple sclerosis. Immunol Rev 144:75-107
3. Steinman L (1996) Multiple sclerosis: a coordinated immunological attack against myelin in the central nervous system. Cell 85:299-302
4. Martin R, McFarland HF (1997) Immunology of multiple sclerosis and experimental allergic encephalomyelitis. In: Raine CS, McFarland HF, Tourtellotte WW (eds) Multiple sclerosis: clinical and pathogenetic basis. Chapman & Hall, London, pp 221-242
5. Ota K, Matsui M, Milford EL, Mackin GA, Weiner HL, Hafler DA (1990) T cell recognition of an immunodominant myelin basic protein epitope in multiple sclerosis. Nature 346:183-187
6. Chou YK, Henderikx P, Vainiene M, Whitham RH, Bourdette DN, Chou CH, Hashim GA, Offner H, Vandenbark AA (1991) Specificity of human T cell clones reactive to immunodominant epitopes of myelin basic protein. J Neurosci Res 28:280-290
7. Martin R, Jaraquemada D, Flerlage M, Richert JR, Whitaker JN, Long EO, McFarlin DE, McFarland HF, Richert J, Whitaker J (1990) Fine specificity and HLA restriction of myelin basic protein-specific cytotoxic T cell lines from multiple sclerosis patients and healthy individuals. J Immunol 145:540-548
8. Zhang J, Medaer R, Hashim GA, Chin Y, van den Berg-Loonen E, Raus JC (1992) Myelin basic protein-specific T lymphocytes in multiple sclerosis and controls: precursor frequency, fine specificity, and cytotoxicity. Ann Neurol 32:330-338
9. Valli A, Sette A, Kappos L, Oseroff C, Sidney J, Miescher G, Hochberger M, Albert ED, Adorini L (1993) Binding of myelin basic protein peptides to human histocompatibility leukocyte antigen class II molecules and their recognition by T cells from multiple sclerosis patients. J Clin Invest 91:616-628
10. Muraro PA, Vergelli M, Kalbus M, Banks D, Nagle JW, Tranquill LR, Nepom GT, Biddison WE, McFarland HF, Martin R (1997) Immunodominance of a low-affinity MHC binding myelin basic protein epitope (residues 111-129) in HLA-DR4 (B1*0401) subjects is associated with a restricted T cell receptor repertoire. J Clin Invest 100:339-349

11. Pette M, Fujita K, Kitze B, Whitaker JN, Albert E, Kappos L, Wekerle H (1990) Myelin basic protein-specific T lymphocyte lines from MS patients and healthy individuals. Neurology 40:1770-1776

12. Lovett-Racke AE, Trotter JL, Lauber J, Perrin PJ, June CH, Racke MK (1998) Decreased dependence of myelin basic protein-reactive T cells on CD28-mediated costimulation in multiple sclerosis patients. J Clin Invest 101:725-730

13. Scholz C, Patton KT, Anderson DE, Freeman GJ, Hafler DA (1998) Expansion of autoreactive T cells in multiple sclerosis is independent of exogenous B7 costimulation. J Immunol 160:1523-1538

14. Wucherpfennig KW, Strominger JL (1995) Molecular mimicry in T cell-mediated autoimmunity: viral peptides activate human T cell clones specific for myelin basic protein. Cell 80:695-705

15. Hemmer B, Fleckenstein BT, Vergelli M, Jung G, McFarland HF, Martin R, Wiesmuller K-H (1997) Identification of high potency microbial and self ligands for a human autoreactive class II-restricted T cell clone. J Exp Med 185:1651-1659

16. Vitetta ES, Berton MT, Burger C, Kepron M, Lee WT, Yin XM (1991) Memory B and T cells. Annu Rev Immunol 9:193-217

17. Matzinger P (1994) Tolerance, danger, and the extended family. Annu Rev Immunol 12:991-1045

18. Swain SL, Croft M, Dubey C, Haynes L, Rogers P, Zhang X, Bradley LM (1996) From naive to memory T cells. Immunol Rev 150:143-167

19. Schweighoffer T, Luce GE, Tanaka Y, Shaw S (1994) Differential expression of integrins alpha 6 and alpha 4 determines pathways in human peripheral CD4+ T cell differentiation. Cell Adhes Commun 2:403-415

20. Horgan KJ, Luce GE, Tanaka Y, Schweighoffer T, Shimizu Y, Sharrow SO, Shaw S (1992) Differential expression of VLA-alpha 4 and VLA-beta 1 discriminates multiple subsets of CD4+CD45R0+ "memory" T cells. J Immunol 149:4082-4087

21. Byrne JA, Butler JL, Cooper MD (1988) Differential activation requirements for virgin and memory T cells. J Immunol 141:3249-3257

22. Croft M, Swain SL (1995) Recently activated naive CD4 T cells can help resting B cells, and can produce sufficient autocrine IL-4 to drive differentiation to secretion of T helper 2-type cytokines. J Immunol 154:4269-4282

23. Dubey C, Croft M, Swain SL (1995) Costimulatory requirements of naive CD4+ T cells. ICAM-1 or B7-1 can costimulate naive CD4 T cell activation but both are required for optimum response. J Immunol 155:45-57

24. Lee WT, Vitetta ES (1991) The differential expression of homing and adhesion molecules on virgin and memory T cells in the mouse. Cell Immunol 132:215-222

25. Bradley LM, Watson SR, Swain SL (1994) Entry of naive CD4 T cells into peripheral lymph nodes requires L-selectin. J Exp Med 180:2401-2406.

26. Bell EB, Sparshott SM (1990) Interconversion of CD45R subsets of CD4 T cells in vivo. Nature 348:163-166

27. Bunce C, Bell EB (1997) CD45RC isoforms define two types of CD4 memory T cells, one of which depends on persisting antigen. J Exp Med 185:767-776

28. Hamann D, Baars PA, Rep MHG, Hooibrink B, Kerkhof-Garde SR, Klein MR, Lier RAW (1997) Phenotypic and functional separation of memory and effector human CD8(+) T cells. J Exp Med 186:1407-1418

29. Smith SH, Brown MH, Rowe D, Callard RE, Beverley PCL (1986) Functional subsets of human help-inducer cells defined by a new monoclonal antibody, UCHL1. Immunology 58:63

30. Akbar AN, Terry L, Timms A, Beverley PCL, Janossy G (1988) Loss of CD45R and gain of UCHL1 reactivity is a feature of primed T cells. J Immunol 140:2171-2178

31. Allegretta M, Nicklas JA, Sriram S, Albertini RJ (1990) T cells responsive to myelin basic protein in patients with multiple sclerosis. Science 247:718-721

32. Zhang J, Markovic-Plese S, Lacet B, Raus J, Weiner HL, Hafler DA (1994) Increased frequency of interleukin 2-responsive T cells specific for myelin basic protein and proteolipid protein in peripheral blood and cerebrospinal fluid of patients with multiple sclerosis. J Exp Med 179:973-984

33. Michie CA, McLean AR, Alcock C, Beverley PCL (1992) Lifespan of human lymphocyte subsets defined by CD45 isoforms. Nature 360:264-265

Chapter 3

From specificity to degeneracy to molecular mimicry: antigen recognition of human autoreactive and pathogen-specific CD4+ T cells

B. HEMMER, C. PINILLA, B. GRAN, H.F. MCFARLAND, R. HOUGHTEN, R. MARTIN

Introduction

Multiple sclerosis (MS) is a chronic demyelinating disease of the central nervous system (CNS). Although the cause of MS has not been identified, it is widely believed that myelin-specific T cells are somehow involved in the disease process [1]. However it is not clear how these cells become activated and mediate CNS disease. One of the most attractive models for the activation of autoreactive T cells is termed the "molecular mimicry hypothesis" [2]. According to this model, autoreactive T cells become activated in the peripheral immune system by crossreactive ligands derived from infectious agents. After activation and upregulation of adhesion molecules, these cells can cross the blood-brain barrier; after reactivation by myelin antigens either directly or via the recruitment of other cells, the activated T cells damage the myelin sheath and oligodendrocytes.

Recently, much has been learned about crossrecognition of T cells. Initially it was believed that T cells could only recognize one or a few ligands. However within the last decade it has become evident that a single T cell receptor (TCR) can interact with many ligands although the stimulatory potency of the ligands may vary significantly [3]. Based on this concept new methods for the identification of crossreactive ligands have been established to prove the molecular mimicry hypothesis at the molecular level [4]. However no methods have been developed to systematically decrypt antigen recognition of T cells and to exploit the extent of degeneracy in T cell antigen recognition. Here we demonstrate the use of synthetic peptide combinatorial libraries in the positional scanning format (PS-SCL) to identify potent ligands for CD4+ class II-restricted autoreactive and foreign antigen-specific T cell clones.

Material and methods

Library and peptide synthesis

The PS-SCL composed of decapeptides was prepared as first presented using the simultaneous multiple peptide synthesis approach (SMPS), methylbenzhydrylamine polystyrene resin, and t-Boc-protected L-amino acids [5]. The solutions were lyophilized and resuspended in water at 10 mg/ml. Individual peptides were synthesized either by SMPS or F-moc technology. The purity and identity of each peptide were characterized by RP-HPLC and MALDI-TOF mass spectrometry.

Each of the 10 positional peptide libraries making up this decapeptide PS-SPCL is composed of 20 peptide mixtures, in which a single position is defined with one of the 20 natural L-amino acids (represented as O), and the remaining nine positions of the 10-residue sequence are composed of mixtures (represented as X) of 19 amino acids (cysteine omitted) (Fig. 1). The 10 positional peptide libraries have N-terminal acetyl and C-terminal amide groups. Each positional library contains the same diversity of peptide sequences; they differ only in the location of their defined position. Theoretically, each peptide mixture is made up of approximately 3×10^{11} (19^9) individual sequences for a library composed of approximately 6×10^{12} decapeptides. Assuming an average molecular weight of 1200 for a peptide mixture and a concentration of

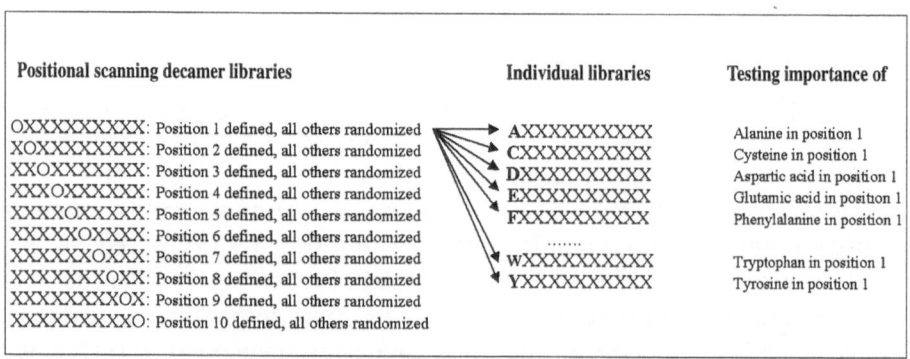

Fig. 1. Combinatorial peptide library design in the positional scanning format. The decamer library used for the T cell assays consists of 10 different positional scanning libraries which contain one defined amino acid in position 1 to 10 (*left side*). Each of these 10 libraries is composed of 20 individual libraries that test the importance of each natural amino acid for the specific position (*right side*, shown for position 1). A total of 200 individual peptide libraries is necessary to test the importance of each amino acid in every individual position of a decamer epitope

10 mg/ml, the concentration of each individual decapeptide is 28 fM [6]. Single letter code for peptides and peptide libraries is used throughout the manuscript.

T cell clones and antigen-presenting cells

T cell clones (TCC) were established from peripheral blood lymphocytes by a limiting dilution, split-well technique with myelin basic protein (MBP) and characterized as described before [7].

T cell proliferation

TCC were rested for 8-12 days, washed and resuspended at 2.5 x10^5 cells/ml in complete medium (CM) composed of IMDM containing 5% human serum, 1% penicillin/streptomycin, and 0.2% gentamicin. One hundred microliters of this cell suspension were added to each well of 96-well U-bottom plates containing 1 x 10^5 irradiated PBMC per well and varying concentrations of peptide libraries or peptides [8]. Cells were cultured for 48 h at 37 °C. During the last 8 h of culture, 1 µCi [^{3}H] -thymidine was added to each well. Cells were then harvested and incorporated radioactivity was measured by scintillation counting. The proliferative response to deduced peptides was measured using conditions described above with varying dilutions of peptides [8]. Each experiment with single peptides was repeated at least twice, each experiment with peptide libraries at least 4 times.

Results

We tested 10 human myelin basic protein (MBP)-specific CD4$^+$ T cell clones for their response to selected mixtures of the decamer PS-SCL. Five TCC showed reproducible proliferative responses to peptide mixtures with defined amino acids. These TCC were tested for their response to all 200 mixtures of the library (10 libraries covering the 20 amino acids in each position) (Fig. 1). For each TCC, at least 5 independent experiments were performed to assure reproducibility of the results. The proliferative responses of the TCC to all libraries for a specific position was compared and the amino acids were ranked according to the stimulatory potency of the corresponding mixture. On the basis of the response to all 200 mixtures, the TCR motif for the individual TCC was established [8-10].

Theoretically these TCR motifs should contain optimal amino acids for the individual positions. Recently we demonstrated that peptides deduced from these motifs, containing optimal amino acids in each individual position, were highly stimulatory for the individual TCC [8]. In addition we were able to define crossreactive antigens from non-self sources that stimulated the TCCs at higher potency than the self-peptide MBP [8]. This observation demonstrated that the PS-SCL approach allow precise determination of TCR motifs and identification of potent peptide ligands.

The fact that 5 of 10 TCC responded to complex mixtures containing individual peptides at concentrations of 0.28 fM and less (PS-SCL concentration 100 μg/ml and less) demonstrated the flexibility of the TCR recognition of those TCC. We assume that millions of different peptides are recognized by those TCC [10]. When comparing the sequence of MBP to the PS-SCL-defined TCR motifs, it became evident that MBP did not align in all positions with the TCR motifs. This observation was true for all human MBP-specific TCC defined by the PS-SCL approach. Similar findings were obtained using single amino acid-substituted peptides [7]. In contrast, a virus-specific CD4+ TCC specific for influenza hemagglutinin peptide (HA(306-318)) had a TCR motif that matched the influenza sequence in all but one position. According to these results it seemed that the HA(306-318) peptide fitted much better to the TCR motif of the foreign antigen-specific TCC than did the MBP-peptide to the recognition motif of the self peptide-specific TCC.

Based on the TCR motifs, we deduced and synthesized peptides for three MBP-specific TCC. In addition we searched protein databases for naturally occurring peptide sequences with the TCR motifs. Based on these motifs we were able to identify high potency peptides for each of the three TCC that stimulated the TCC at picomolar concentrations. For 2 of the TCC we defined peptides from self and non-self peptides that stimulated the TCC, some even more potently than the MBP peptide ([8], and B. Hemmer et al., submitted). Some of those peptides were derived from viral proteins such as cytomegalovirus and human herpesvirus 7 ([8], and B. Hemmer et al., submitted). In addition we identified a peptide from myelin-oligodendrocyte protein as well as 5-residue peptides that were recognized by one of the MBP-specific TCC with low affinity ([8], and B. Hemmer et al., submitted).

Discussion

Our previous results and data reported in this article demonstrate that human T cells can respond to mixtures of $6x10^{12}$ peptides. Since this was observed with several autoreactive and one virus-specific TCC, it can be assumed that recognition of highly diverse peptide mixtures is rather a common feature of T cells than the exception. The fact that T cells can respond to such complex mixtures of peptides indicates a tremendous flexibility or degeneracy in antigen recognition. Recent studies on T cell selection in the thymus and survival of T cells in the periphery have established a role of TCR-mediated signals for those processes [11, 12]. Since T cells need a continuous stimulus through their TCR, it seems likely that this is achieved by self-MHC loaded with self peptides. Although the repertoire of TCR ligands in the thymus and the peripheral lymphatic system is limited, a heterogeneous repertoire of different T cells has to be selected and stimulated by those complexes. Therefore, flexibility in TCR antigen recognition seems the only solution for this obvious contradiction [10]. According to our results millions of peptides should be stimulatory for a given TCR. If the threshold for positive selection and survival of T cells in the periphery is low enough, all those peptides or peptide mixtures should support continuous TCR ligation. This not only would explain why a single peptide can select a heterogeneous TCR repertoire but also why the same TCR can be selected by very different peptides. The same mechanisms should act in the process of T cell survival in the periphery. It seems possible that the low but continuous engagement of the TCR is a result of TCR degeneracy.

In addition to these implications for T cell survival, our findings have important consequences for the understanding of autoimmune responses. Since T cells can recognize a variety of ligands it seems likely that cross-reactive T cells become activated during the response to infectious agents. The finding that all MBP-specific TCC recognized the self peptide as a suboptimal ligand whereas the foreign-antigen specific TCC recognized the HA peptide almost as an optimal ligand indicates substantial differences in the affinity of those ligands for the different TCR. It seems that the self antigen-specific TCC recognize the self peptide with low affinity whereas the foreign antigen-specific TCC recognizes its antigen with high affinity. This would suggest that the non-self specific T cells found in the periphery are the result of in vivo selection during infection. In agreement with recent findings it seems that an infection selects only a few T cells carrying optimal TCR for the particular peptide ligand. Since optimal fit was not observed with MBP-specific TCC using the PS-SCL [8] nor the peptide analogue approach [7, 13], we hypothesize that those T cells were not expanded by MBP but by a so far unknown infectious agent. The initial selection process involved a high affinity interaction and selection for the non-self ligand (Fig. 2). However after activation of the T cells and upregulation of adhesion molecules these T cells may have entered the brain. In the activated state those T cells can crossreact with the self MBP antigen and thus mediate inflammation and damage to the brain. The hypothesis of foreign antigens being the high affinity ligands and self antigens being the low affinity ligands for the same TCR adds a new twist to the molecular mimicry hypothesis that incorporates recent findings on T cell responses during infectious diseases.

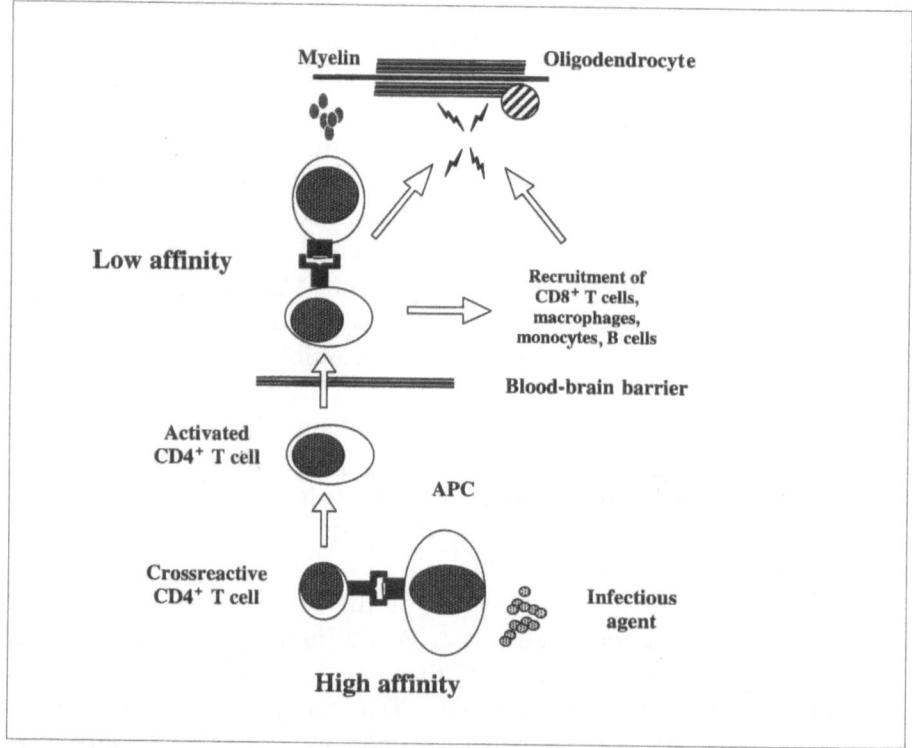

Fig. 2. Molecular mimicry. T cells become activated in the peripheral lymphoid tissue during infectious diseases. The infectious agent selects for high affinity TCRs specific for the particular peptide ligands. Activated T cells can cross the blood-brain barrier and some of them might crossrecognize self-antigens in the brain with low affinity. Reactivation of these T cells can damage oligodendrocytes and the myelin sheath either directly or via the recruitment of other cells. *APC,* antigen-presenting cell

Summary

CD4$^+$ T cells recognize peptides bound to major histocompatibility complex (MHC) molecules through interaction with their T cell receptor (TCR). Initially, this interaction was considered to be highly specific. However, recently it became evident that a specific TCR can interact with many ligands. We have defined recognition patterns of a set of myelin basic protein (MBP)-specific and influenza-virus specific human T cell clones (TCC) using soluble synthetic peptide combinatorial libraries in the positional scanning format (PS-SCL). The use of PS-SCL allowed determination of important residues for each position of the antigenic peptide and deduction of recognition motifs for each individual TCC. All MBP-specific TCCs had motifs that identified MBP as a rather suboptimal antigen. In addition, based on the recognition motifs, high potency ligands were found for the MBP-specific TCC, some of them derived from viral antigens such as human herpesvirus 7. For one TCC, we identified a crossreactive peptide derived from myelin oligo-dendrocyte glycoprotein as well as 5-residue synthetic peptides that were more potent than the autoantigen used to establish the TCC. These data demonstrate that antigen recognition of CD4$^+$ T cells is highly degenerate, allowing positive engagement by many ligands. Although initial activation of these cells in vivo will most likely require a high affinity interaction (such as that of a viral antigen), the activated cells, after upregulation and activation of coreceptors, have a high potential for cross-recognition of self antigens. These results shed new light on the extent of cross-recognition in the immune system and its impor-tance for the pathogenesis of autoimmune diseases.

Acknowledgement
This study was supported in part by a Fogarty fellowship and a grant of the Deutsche Forschungsgemeinschaft (He 2386/2-1).

References

1. Martin R, McFarland HF, McFarlin DE (1992) Immunological aspects of demyelinating diseases. Annu Rev Immunol 10:153-187
2. Fujinami RS, Oldstone MBA (1985) Amino acid homology between the encephalitogenic site of myelin basic protein and virus: mechanism for autoimmunity. Science 230:1043-1045
3. Evavold BD, Sloan-Lancaster J, Wilson KJ, Rothbard JB, Allen PM (1995)

Specific T cell recognition of minimally homologous peptides: evidence for endogenous ligands. Immunity 2:655-663

4. Wucherpfennig KW, Strominger JL (1995) Molecular mimicry in T cell-mediated autoimmunity: viral peptides activate human T cell clones specific for myelin basic protein. Cell 80:695-705

5. Houghten RA (1985) General method for the rapid solid-phase synthesis of large numbers of peptides: specificity of antigen-antibody interaction at the level of individual amino acids. Proc Natl Acad Sci USA 82:5131-5135

6. Pinilla C, Appel JR, Houghten RA (1994) Investigation of antigen-antibody interactions using a soluble nonsupport-bound synthetic decapeptide library composed of four trillion sequences. Biochem J 301:847-853

7. Vergelli M, Hemmer B, Kalbus M, Vogt AB, Ling N, Conlon P, Coligan JE, McFarland H, Martin R (1997) Modifications of peptide ligands enhancing T cell responsiveness imply large numbers of stimulatory ligands for autoreactive T cells. J Immunol 158:3746-3752

8. Hemmer B, Fleckenstein BT, Vergelli M, Jung G, McFarland H, Martin R, Wiesmuller KH (1997) Identification of high potency microbial and self ligands for a human autoreactive class II-restricted T cell clone. J Exp Med 185:1651-1659

9. Hemmer B, Pinilla C, Appel J, Pascal J, Houghten R, Martin R (1998) The use of soluble synthetic peptide combinatorial libraries to determine antigen recognition of T cells. J Pept Res 52:338-345

10. Hemmer B, Vergelli M, Pinilla C, Houghten R, Martin R (1998) Probing degeneracy in T cell recognition using peptide combinatorial libraries. Immunol Today 19:163-168

11. Kisielow P, von Boehmer H (1995) Development and selection of T cells: facts and puzzles. Adv Immunol 58:87-209

12. Kirberg J, Berns A, von Boehmer H (1997) Peripheral T cell survival requires continual ligation of the T cell receptor to major histocompatibility complex-encoded molecules. J Exp Med 186:1269-1275

13. Hemmer B, Vergelli M, Gran B, Ling N, Conlon P, Pinilla C, Houghten R, McFarland HF, Martin R (1998) Cutting edge: Predictable TCR antigen recognition based on peptide scans leads to the identification of agonist ligands with no sequence homology. J Immunol 160:3631-3636

Immune activation in the interface between innate immunity and adaptive response: in vitro studies and therapeutical implications

G. Ristori, A. Perna, C. Montesperelli, L. Battistini, R. Bomprezzi, S. Cannoni, G. Borsellino, C. Pozzilli, C. Buttinelli, M. Salvetti

Introduction

In multiple sclerosis (MS), as well as in other putatively T cell mediated diseases, much attention has been devoted to possible dysregulations at the level of the adaptive immune response, with particular emphasis on the search for potential T lymphocyte autoantigens. These investigations have been, in many cases, unrewarding. One reason may be that T cells are not the key players in self-nonself discrimination: T cell receptors (TCR) have randomly generated specificities that, for this reason, cannot determine the origin and biological context of their ligands [1]. It is becoming increasingly evident that this task is better carried out by the innate immune system that relies on germline-encoded receptors which bind to invariant molecules shared by large groups of microorganisms. These structures (techoic acids, lipopolysaccharides (LPS), double-stranded RNA, mannans) are therefore readily recognized as markers of infection by the innate immune system that instructs the adaptive response accordingly [2]. Based on this evidence, it is now clear that the correct functioning of innate and adaptive immunity as well as their balanced interaction are essential for a physiological immune response [3]. This has prompted much of the recent research focused on the interface between these two arms of the immune system in autoimmunity.

Serum amyloid A protein

Our first approach to the problem in MS was to study serum amyloid A protein (SAA). Together with C-reactive protein (CRP), SAA is one of the major acute phase reactants. It is synthesized by hepatocytes in response to cytokines and other regulatory factors [4], it binds extracellular matrix glycoproteins and can induce adhesion and migration of CD4+

and CD8[+] T cells, monocytes and macrophages [5, 6]. Besides its speculative interest as a marker of activation of the innate immune system, it is a potential candidate as an indicator of ongoing inflammation in the peripheral immune compartment and may become useful as a surrogate marker of disease activity.

We performed serial monitoring of SAA in the peripheral blood of patients with relapsing-remitting MS over a 3-month period [7]. Patients were monitored in parallel with gadolinium-enhanced magnetic resonance imaging (Gd-MRI) of the brain. The results showed that signs of ongoing peripheral inflammation, reflected by increased SAA levels, could be detected in MS patients (Table 1). Since SAA has a role in the interplay between innate immunity and acquired immune response [5, 6], it can be speculated that, in MS, proinflammatory influences from the innate immune system contribute to the activation of the autoimmune repertoire in the periphery. The increased frequency of activated myelin-specific T cells in patients with MS is in line with this possibility [8, 9].

Table 1. Monthly (I-IV) SAA measurements, with number and volume of enhancing areas at monthly (I-IV) Gd-MRI

Patient	SAA values (µg/ml)				Number (volume)[a] of enhancing areas at Gd-MRI			
	I	II	III	IV	I	II	III	IV
1	25	31	27	19	2 (81)	3 (295)	11 (567)	14 (690)
2	27	25	12	11	1 (332)	0[b]	0[b]	0[b]
3	29	17	12	11	0[b]	0[b]	1 (317)	0[b]
4	15	22	16	12	0[b]	1 (37)	2 (144)	2 (114)
5	11	11	15	8	4 (140)	3 (173)	1 (48)	0[b]
6	10	9	9	8	9 (1337)	8 (1436)	12 (1883)	9 (959)
7	14	3[b]	18	13	4 (243)	6 (583)	3 (620)	6 (900)
8	11	4[b]	9	8	1 (48)	0[b]	0[b]	0[b]
9	29	5	<2[b]	<2[b]	2 (144)	1 (74)	4 (376)	3 (757)
10	11	<2[b]	5	<2[b]	1 (129)	2 (790)	2 (232)	2 (125)
11	3[b]	4[b]	4[b]	<2[b]	1 (232)	1 (203)	1 (255)	4 (431)

[a] Total volume in mm^3
[b] Normal SAA levels; no enhancing lesions

γδ T lymphocytes

Among the cells that operate at the boundaries between innate and adaptive immunity, γδ T lymphocytes have received much interest given their possible involvement in various autoimmune processes. γδ T lymphocytes are present in MS plaques as well as in the cerebrospinal fluid as two distinct subsets, suggesting specific roles of these lymphocytes in disease pathogenesis [10, 11]. Furthermore, these cells appear to colocalize with autoantigens (60-kDa heat shock proteins) they may recognize in MS lesions [12]. Investigations on the TCR usage of γδ lymphocytes in plaques have shown a predominant presence of a Vδ2-Jδ3 rearrangement that harbors a relatively invariant and germline-encoded CDR3 sequence [13]. Interestingly, the same sequence was subsequently found in T cell lines specific for another class of potential autoantigens in this disease, namely the 70-kDa heat shock proteins. It is possible that, in MS plaques, clonal expansion of γδ T cells takes place in response to this class of heat shock proteins. Such responses may be relevant since γδ T cells possess considerable cytotoxic activity, also against oligodendrocytes [14], as well as the capacity to produce cytokines (in particular interferon-γ) and chemokines involved in the recruitment of monocytes/macrophages [10]. But the interest for γδ T cells as elements that operate within a system at limited polymorphism stems not only from the invariant sequences found in some TCRs but also from the demonstrated ability of these cells to respond to non-proteinaceous moieties of microbial origin. A possibility that remains to be assessed is the pathogenetic potential of γδ cells that may cross-recognize self and foreign non-proteinaceous molecules. The evidence collected so far in this respect is highly speculative since it is inferred from the upregulated expression of CD1b in MS lesions (the CD1 lineage of antigen-presenting molecules appears to have evolved to bind and present non-protein lipid and glycolipid self and foreign antigens) [15]. A final aspect that links γδ T cells to the innate immune system is the presence, in a portion of these lymphocytes, of surface markers that are regulatory molecules characteristic of NK cells (CD16, CD56 and different inhibitory NK cell receptors (NKR) for HLA class I molecules). This suggests some common regulatory pathways between NK cells and γδ cells. In γδ lymphocytes one of these molecules, NKRP1A (CD161), is expressed almost exclusively by the Vδ2 subset [16, 17]. The percentage of circulating Vδ2+ γδ T cells coexpressing NKRP1A is significantly increased in patients with MS compared with healthy donors, and the expression of

NKRP1A on the cell surface of γδ cells correlates with an increased ability to transmigrate across the vascular endothelium [18].

N-formylated peptide reactivity

A different aspect, again in between the adaptive and the innate responses is the reactivity to N-formylated peptides. N-formylated peptides are typical products of prokaryotes and mitochondria since both initiate protein synthesis with an N-formylated methionine. In eukaryotes, the initiator methionine is not formylated [19]. Therefore N-formylated peptides can be readily recognized by TCR as markers of infection, making the TCR more akin to a germline-encoded receptor of the innate immune system. This system has received much attention in rodents, where N-formylated peptides are presented to CD8 T cells in the context of a class I-like molecule with low polymorphism, H2-M3, therefore reinforcing the idea of a response resembling those of innate immunity [20]. A T cell response to N-formylated peptides has now been demonstrated in humans (G. Ristori et al., manuscript in preparation), paving the way to studies on the T cell response to N-formylated peptides in human autoimmune diseases in which autoreactivity to N-formylated peptides of mitochondrial origin may occur.

This issue was addressed in a study in which peripheral blood T cells from 31 patients with MS and 31 age and sex-matched healthy controls were assayed for their proliferative responses to a panel of:
- N-terminal nonamers of COXII, ATP8, ND4, CYTB human mitochondrial proteins, 60- and 70-kDa heat shock proteins, and 32-kDa secreted antigen of *M. tuberculosis* (kindly provided by Prof. Alberto Chersi, Rome, Italy);
- Submitochondrial particles prepared by sonication of the mitochondrial suspension and separation of the water-soluble fraction (kindly provided by Prof. Paolo Riccio, Potenza, Italy) (G. Ristori et al., unpublished).

At the present stage, no significant differences have been observed between patients and controls. Nonetheless, the T cell response to the N-termini of the remaining mitochondrial proteins needs to be assessed before definitive conclusions can be drawn about the T cell responses to N-formylated peptides of mitochondrial origin in MS.

Therapeutical implications

If innate immunity shows signs of dysregulation, how is it possible to operate on the innate-adaptive immunity axis in order to restore protective responses and eliminate the harmful ones? Exposure to microorganisms may do this, and there is epidemiological evidence that the so-called Westernization, with its decreased microbial exposure, may have favored the rise in incidence and prevalence of atopic and autoimmune diseases [3, 21, 22].

Following the results of recent experiments in the field of insulin-dependent diabetes mellitus (IDDM) [23, 24], we performed an exploratory trial of adjuvant therapy (bacille Calmette-Guérin (BCG) vaccination in MS). The study was performed according to published guidelines [25] and was a single crossover, magnetic resonance (MR)-monitored trial (monthly follow-up with gadolinium (Gd)-enhanced MR scans of the brain for 6 months of run-in and for 6 months after BCG vaccine) in 14 patients with relapsing-remitting MS. Treatment efficacy was assessed by comparing MR-detected disease activity before and after BCG vaccine. Two patients dropped out for events not related to treatment. The analysis was therefore performed on 12 cases. We observed 9 clinical relapses before BCG and 3 after. No adverse effects were reported except for local reaction to inoculation in 2 patients. After treatment, MR-detected disease activity was significantly reduced (51% and 57% reduction in Gd-enhancing and in active lesions, respectively; $p = 0.008$ by Wilcoxon signed rank test). Accordingly, the mean number of Gd-enhancing and active scans significantly decreased after BCG vaccine (0.42% vs. 0.27%, $p = 0.04$; 0.64% vs. 0.37%, $p = 0.02$, respectively).

Conclusions

These data support the view that the interaction between innate and adaptive immunity may harbor some of the critical steps in the pathogenesis of MS and may offer an opportunity for therapeutic interventions.

References

1. Medzhitov R, Janeway CA Jr (1997) Innate immunity: the virtues of a non-clonal system of recognition. Cell 91:295-298
2. Fearon DT, Locksley RM (1996) The instructive role of innate immunity in the acquired immune response. Science 272:50-53
3. Rook GA, Stanford JL (1998) Give us this day our daily germs. Immunol Today 19:113-116
4. Steel DM, Whitehead AS (1994) The major acute phase reactants: C-reactive protein, serum amyloid P component and serum amyloid A protein. Immunol Today 15:81-88
5. Xu L, Badolato R, Murphy WJ, Longo DL, Anver M, Hale S, Oppenheim JJ, Wang JM (1995) A novel biologic function of serum amyloid A (SAA). Induction of T lymphocyte migration and adhesion. J Immunol 155:1184-1190
6. Preciado-Patt L, Hershkovitz R, Fridkin M, Lider O (1996) Serum amyloid A binds specific extracellular matrix glycoproteins and induces adhesion of resting CD4 T cells. J Immunol 156:1198-1205
7. Ristori G, Laurenti F, Stacchini P, Gasperini C, Buttinelli C, Pozzilli C, Salvetti M (1998) Serum amyloid A protein is elevated in relapsing-remitting multiple sclerosis. J Neuroimmunol 88:9-12
8. Zhang J, Markovic-Plese S, Lacet B, Raus J, Weiner HL, Hafler DA (1994) Increased frequency of interleukin 2-responsive T cells specific for myelin basic protein and proteolipid protein in peripheral blood and cerebrospinal fluid of patients with multiple sclerosis. J Exp Med 179:973-984
9. Allegretta M, Nicklas JA, Sriram S, Albertini RJ (1990) T cells responsive to myelin basic protein in patients with multiple sclerosis. Science 247:718-721
10. Wucherpfennig KW, Newcombe J, Li H, Keddy C, Cuzner ML, Hafler DA (1992) γδ T cell receptor in acute multiple sclerosis lesions. Proc Natl Acad Sci USA 89:4588-4592
11. Triebel F, Hercend T (1989) Subpopulations of human peripheral γδ T lymphocytes. Immunol Today 10:186-188
12. Selmaj K, Brosnan CF, Raine CS (1991) Co-localization of lymphocytes bearing γδ T cell receptor and heat shock protein hsp65$^+$ oligodendrocytes in multiple sclerosis. Proc Natl Acad Sci USA 88:6452-6456
13. Battistini L, Selmaj K, Kowal C, Ohmen J, Modlin RL, Raine CS, Brosnan CF (1995) Multiple sclerosis: limited diversity of the Vδ2-Jδ3 T cell receptor in chronic active lesions. Ann Neurol 37:198-203
14. Freedman MS, Ruijs TCG, Selin LK, Antel JP (1991) Peripheral blood γ-δ T cells lyse fresh human brain-derived oligodendrocytes. Ann Neurol 30:794-800
15. Battistini L, Fisher F, Raine CS, Brosnan CF (1996) CD1b expression in multiple sclerosis. J Neuroimmunol 67:145-151

16. Battistini L, Borsellino G, Sawicki G, Poccia F, Salvetti M, Ristori G, Brosnan CF (1997) Phenotypic and cytokine analysis of human peripheral blood γδ T cells expressing NK cell receptor. J Immunol 159:3723-3730

17. Poggi A, Zocchi MR, Costa P, Ferrero E, Borsellino G, Placido R, Galgani S, Salvetti M, Gasperini C, Ristori G, Brosnan CF, Battistini L (1999) IL-12-mediated NKRP1A up-regulation and consequent enhancement of endothelial transmigration of Vδ2+ TCRγδ+ T lymphocytes from healthy donors and multiple sclerosis patients. J Immunol (*in press*)

18. Poggi A, Costa P, Zocchi MR, Moretta L (1997) Phenotypic and functional analysis of CD4+ NKRP1A human T lymphocytes: direct evidence that the NKRP1A molecule is involved in transendothelial migration. Eur J Immunol 27:2345-2350

19. Lewin B (1993) Genes V. Oxford University, Oxford

20. Fischer Lindahl K, Byers DE, Dahbi VM, Hovik R, Jones EP, Smith GP, Wang CR, Xiao H, Yoshino M (1997) H2-M3, a full-service class Ib histocompatibility antigen. Annu Rev Immunol 15:851-879

21. Shirakawa T, Enomoto T, Shimazu S, Hopkin JM (1997) The inverse association between tuberculin response and atopic disorder. Science 275:77-79

22. Ristori G, Buttinelli C, Pozzilli C, Fieschi C, Salvetti M (1999) Microbe exposure, innate immunity and autoimmunity. Immunol Today (*in press*)

23. Shehadeh N, Calcinaro F, Bradley BJ, Bruchlim I, Vardi P, Lafferty KJ (1994) Effect of adjuvant therapy on development of diabetes in mouse and man. Lancet 343:706-707

24. Pozzilli P, on behalf of the IMDIAB Group (1997) BCG vaccine in insulin-dependent diabetes mellitus. Lancet 349:1520-1521

25. Miller DH, Alpert PS, Barkhof F, Frances G, Frank JA, Hodgkinson S, Lublin FD, Paty DW, Reingold SC, Simon J (1996) Guidelines for the use of magnetic resonance techniques in monitoring the treatment of multiple sclerosis. US National MS Society Task Force. Ann Neurol 39:6-16

Neuronal control of the immunological microenvironment in the CNS: implications on neuronal cell death and survival

H. NEUMANN, T. MISGELD, I. MEDANA

Neuronal signals modulate the immunological microenvironment

Transection paradigms in the nervous system

The central nervous system (CNS) has an immunoprivileged status. In the healthy CNS, class I as well as class II major histocompatibility (MHC) molecules are virtually absent. Heterodimeric MHC molecules are essential for the initiation, propagation and effector phases of antigen-specific immune responses. Endogenous and exogenous antigenic peptides are presented via MHC molecules to T lymphocytes to enable cognate interactions. While MHC molecules are absent in the intact CNS, they are inducible on different brain cell types during inflammatory or neurodegenerative diseases. Recent evidence for the involvement of neurons in the regulation of MHC expression emerged from several studies using neuronal transection models. These models permit the analysis of cellular responses occurring locally, as well as those distant from the primary lesion, without interfering with the blood-brain barrier.

Transection of axons in the CNS induces a local immune response with expression of MHC molecules. Furthermore, immune activation occurs in regions distant from the lesion (Fig. 1). *De novo* expression of MHC class I and class II molecules has been detected in the facial nucleus on glial cells surrounding the perikarya of motoneurons following peripheral nerve transection [1]. Even more surprising than retrograde immune activation is the observed change in the immunological microenvironment in the denervated area. In particular, axonal transection of mossy fibers in the rat hippocampus, which abrogates normal physiological input to the innervated target tissue, dramatically

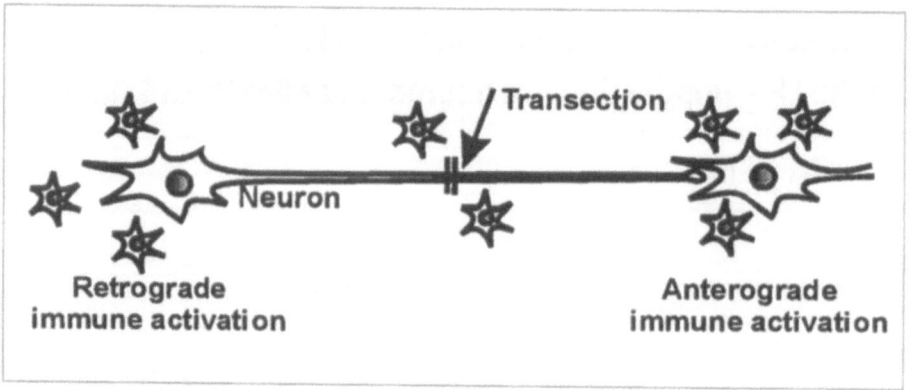

Fig. 1. Retrograde and anterograde immune activation following an axonal lesion. Axonal transection in the CNS leads to MHC expression on glial cells nearby the neuronal perikarya (retrograde immune activation) and in the innervated target tissue (anterograde immune activation)

enhances expression of MHC class II molecules in the innervated hippocampal formation [2].

Neuronal lesions also stimulate homing of T lymphocytes to the lesioned neuronal perikarya [3, 4]. Transection of the facial nerve in mice induces retrograde infiltration of T lymphocytes in the facial nucleus peaking 14 days after motoneuron lesion. This is surprising since the blood-brain barrier is not disturbed in this area of immune activation. Accumulation of T lymphocytes juxtaposed to lesioned neurons is prominent after transfer of in vitro activated T lymphocytes. These activated T lymphocytes can be found interacting with the lesioned neurons (Fig. 2). Homing is also observed when T lymphocytes specific for myelin basic protein are injected intravenously in rats which had previously received a motoneuron [4] or optic nerve lesion [5]. Within hours, T lymphocytes from the blood stream find their way through the intact blood-brain barrier to the neuronal somata of the peripherally lesioned nerves.

Recently, we observed an additional feature of immune activation in response to neuronal lesion. Proinflammatory cytokines are locally produced in the CNS in response to a distant peripheral nerve lesion. Peripheral facial nerve transection induces, retrogradly, gene transcription of tumor necrosis factor-α (TNF-α), interleukin-1β (IL-1β) and interferon-γ (IFN-γ) in the facial nucleus [3]. The principal source of these proinflammatory cytokines could be activated microglia or infiltrating immune cells. In several studies, neurons have been demonstrat-

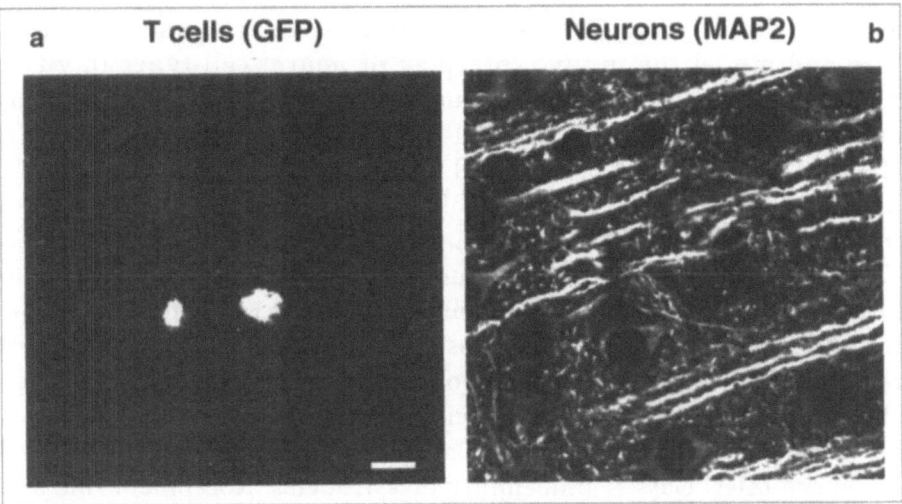

Fig. 2a, b. *Activated auto-reactive T lymphocytes infiltrating brain tissue.* Green fluorescent protein (GFP)-transduced T lymphocytes (a) were found in close proximity to neurons immunolabeled with MAP2 (b) and detected by confocal laser scanning microscopy. Scale bar, 10 µm

ed to express cytokines like TNF-α and IL-1 [6] or even IFN-γ [7]. Whether neurons secrete proinflammatory cytokines in sufficient amounts to modulate the immune microenvironment remains to be shown.

Taken together, these transection studies established that disruption of neuronal integrity not only initiates local, but also anterograde and retrograde expression of "immunologically-relevant" molecules such as MHC. Consequently, neurons as well as neighboring cells would become competent for interactions with T lymphocytes, which patrol the nervous system parenchyma.

Electrically active neurons control MHC expression

While these in vivo studies described an influence of neurons on the immune status of nervous tissue, little is known about the specific neuronal signals that initiate and terminate this process. Therefore, a number of in vitro studies were performed in our laboratory to clarify this question. It was found that MHC expression, a prerequisite for antigen presentation, is under strict negative control of functionally intact neurons [8].

MHC expression was examined on glial cells in a hippocampal

explant culture system. These explants preserve their organotypic structure and model the in vivo interplay of neural cell types in vitro.
Analysis of this cultured brain tissue demonstrated that intact neurons
prevent induction of MHC class II molecules on astrocytes [8]. MHC
class II expression was not induced on astrocytes neighboring neurons
in untreated explants. However, blockade of neuronal activity with
tetrodotoxin (TTX) restored MHC class II inducibility on astrocytes.
The release of neurotransmitters, such as glutamate, is directly linked to
neuronal electrical activity, and neurotransmitters can act on neighboring astrocytes by inducing Ca^{2+} oscillations via specific ligand-gated
receptors [9]. Glutamate (at high concentrations, 0.1–10 mM) has been
shown to affect MHC class II-inducibility of astrocytes [10]. Other classic neurotransmitters and neuropeptides have also been demonstrated
to reduce MHC class II inducibility of astrocytes. Norepinephrine and
vasoactive intestinal polypeptide (VIP) seem to act directly on cultured
astrocytes preventing IFN-γ-mediated induction of MHC class II molecules [11, 12].

Within hippocampal tissue explants, intact neurons also down-regulat-
ed the inducibility of MHC class II molecules on neighboring microglia
[8]. MHC class II molecules of microglia were stimulated in these cultured
brain slices by IFN-γ. Additional blockade of neuronal activity with TTX
significantly increased the expression of MHC class II on microglia.
Recent data indicate that glutamate released by electrically active neurons was indirectly involved in this inhibitory process and prevented
MHC class II induction on microglia [13]. In our laboratory, glutamate
(at lower concentrations, 20 μM) was found to reduce MHC class II
inducibility of microglia in cultured brain tissue [13]. However, no direct
effect of glutamate on isolated microglial cells was observed.
Production and secretion of neurotrophins are also closely associated
with electrical activity of neurons [14]. In our brain tissue culture
model, nerve growth factor (NGF), brain-derived growth factor (BDNF)
and neurotrophin-3 (NT-3) were independently capable of downregulat-
ing IFN-γ-mediated expression of MHC class II on microglia [13]. NGF
and NT-3, but not BDNF, directly acted on microglia cells isolated in culture and inhibited MHC class II expression via the p75 neurotrophin
receptor [13]. BDNF probably was having an indirect effect on
microglial MHC class II expression in tissue culture. This effect might
have been mediated by neurons releasing NGF in response to stimulation by BDNF.

Neurons not only controlled MHC expression of neighboring glial
cells, but also their own expression of MHC class I molecules [15]. MHC

class I gene transcripts were analyzed in dissociated hippocampal neurons by combining patch-clamp electrophysiology and single cell reverse transcriptase polymerase chain reaction (RT-PCR). Gene transcripts identical to classical MHC class I and β2-microglobulin were detected in IFN-γ-treated neurons [15]. Blockade of neuronal activity with TTX significantly increased the percentage of cultured neurons transcribing MHC class I genes. Immunohistochemistry showed no cell membrane expression of MHC class I in untreated neurons. IFN-γ induced MHC class I expression on a small subpopulation of neurons (Fig. 3). Inducibility of MHC class I expression on the cell membrane was increased after blockade of neuronal activity with TTX and most neurons showed MHC class I molecules on the cell surface. Again, glutamate was involved in MHC regulation, as addition of this neurotransmitter to TTX-treated neuronal cultures reduced their inducibility for MHC I expression [16].

Recently, MHC class I gene transcripts were detected in neurons in the cat visual system by in situ hybridization [17]. The hybridization signal was decreased in the lateral geniculate nucleus following post-natal monocular activity blockade with TTX [17]. In situ hybridization, however, does not allow discrimination between classical and the highly homologous non-classical MHC class I gene transcripts. For example, the observed hybridization signal could be explained by neuronal expression of the non-classical MHC class I molecule HLA-E/ Qa-1 possibly protecting neurons against natural killer cell lysis.

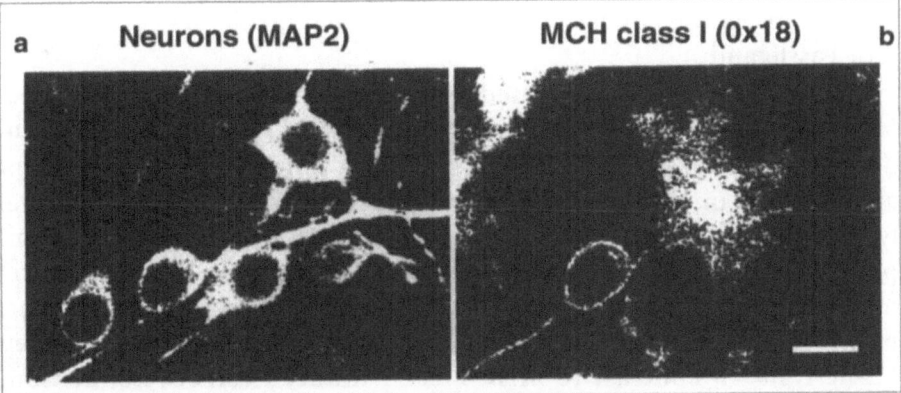

Fig. 3a, b. *Induction of MHC class I molecules on cultured rat neurons.* Neurons treated with 100 U/ml IFN-γ for 72 h were immunolabeled with the neuronal cytoskeleton protein MAP2 (**a**) and antibodies directed against MHC class I (Ox18) (**b**). Only 1 out of 5 neurons showed MHC class I molecules on the cell surface. Scale bar, 10 μm

Role of immune cells in neuronal cell death and survival

The destructive effects of immune cells on neurons

The destructive potential of immune cells has been implicated in numerous neurodegenerative processes [18]. Recently, evidence for a specific cellular immune response in paraneoplastic cerebellar degeneration has been presented [19]. MHC class I-restricted cytotoxic T lymphocytes directed against the cerebellar auto-antigen cdr2 were expanded in patients with paraneoplastic cerebellar degeneration. In addition, natural killer cells have been demonstrated to promote neuronal degeneration in a model of chemically-induced neuronal lesions [20]. Although both studies did not address whether the immune cells directly interact with neurons, their results suggest that leukocytes could be involved in the attack of neurons.

Furthermore, monocytes, which invade lesioned nervous tissue, and locally activated microglia promote neuronal death in acute CNS injury. In particular, macrophages have been shown to release nitric oxide (NO) and glutamate, which exerted a toxic action on co-cultured neurons [21]. Recently invading macrophages have been shown to induce apoptosis in developing neurons by releasing NGF, which acts via the neuronal p75 neurotrophin receptor [22]. The p75 neurotrophin receptor is a member of the TNF receptor family, important receptors for apoptosis signaling. Neurons also express other members of the TNF receptor family. Neurons have been demonstrated to express the p55 TNF receptor [23] and inducible Fas molecules [24]. Therefore, apoptosis could possibly be induced in neurons by immune system-derived mediators such as TNF-α or Fas-ligand.

We and others have reported that neurons are induced to express MHC class I molecules following IFN-γ treatment [15, 24]. This may render neurons susceptible to cytotoxic attack during viral infections or autoimmune or neurodegenerative diseases. Recently our group analyzed mechanisms of neuronal cytotoxicity following cytotoxic T lymphocyte attack. Continuous intracellular calcium ($[Ca^{2+}]_i$) measurements of neurons were performed at the single cell level to discriminate between Fas-mediated apoptosis and perforin-mediated plasma membrane lysis. In contrast to MHC class I-induced astrocytes, neurons attacked by allogenic cytotoxic T lymphocytes did not show typical perforin-mediated, early $[Ca^{2+}]_i$ responses but, rather, gradual and sustained increase in $[Ca^{2+}]_i$ suggestive of apoptosis (I. Medana, unpublished observations). In addition, features of apoptosis are detectable in neurons following attack by cytotoxic T lymphocytes. These results suggest that neuronal cell death can be induced in susceptible neurons by cytotoxic T lymphocytes via Fas-mediated apoptosis.

Beneficial effects of the immune response on neurons

The immune system, however, does not only damage CNS tissue. For example, local transfer of activated immune cells as well as autoimmune infiltrates can promote the regeneration of severed central axons. It has recently been demonstrated that grafting cultured microglial cells into the lesioned spinal cord of adult rats enhances neurite outgrowth [25]. Furthermore, implantation of macrophages preexposed to peripheral nerve segments stimulated tissue repair and motor function of transected rat spinal cord [26]. Also, autoreactive T lymphocytes protected lesioned CNS neurons from secondary degeneration. Rats injected with myelin basic protein-specific T lymphocytes retained more retinal ganglion cells than did control rats after optic nerve injury [27]. The underlying mechanisms of this growth-promoting potential of inflammatory cells is still a matter of debate.

Recent data indicate that proinflammatory cytokines produced by invading immune cells can support CNS repair and neuronal survival. IL-1β and TNF-α can induce glial cells to secrete neurotrophic factors, such as NGF, which are essential mediators for regeneration of nervous tissue [28,29]. Proinflammatory cytokines appear to act directly on neurons and to have neuroprotective effects. TNF-α, for example, protects neurons against cell death induced by ischemia [23]. Mice with targeted mutations of the TNF-receptors, p55 and p75, display exacerbated CNS tissue damage compared with wild type animals, following an ischemic insult [30]. A beneficial effect of TNF-α on neurons was also demonstrated in the injured optic nerve of rabbits [31]. Local application of TNF-α resulted in outgrowth of axons that traversed the injured site. Taken together, these observations provide evidence that immune cell-derived cytokines are capable of supporting neuronal survival and regeneration.

Another potential role for immune cells in neuronal survival is suggested by studies showing that immune cells could produce neurotrophic factors. NGF has been demonstrated to be secreted by T lymphocytes [32] and macrophages [33]. Similarly, these cells produce significant amounts of BDNF [34]. However, the in vivo relevance of neurotrophin secretion by immune cells remains unclear.

Conclusions

Our understanding of the molecular mechanisms that regulate the local immune status of the CNS is by far incomplete. Recently it became apparent that neurons are of central importance to this process, with intact neuronal innervation being essential for the maintenance of the immune privileged status of the CNS. Electrically active neurons reduce MHC inducibility on themselves as well as on neighboring glial cells. However, as soon as the innervation is altered, immune responsiveness increases. The blockade of neuronal activity subsequently enables the induction of MHC molecules on neurons and glial cells by IFN-γ. By regulating immune responsiveness, neurons can control the long-term integrity of their surrounding tissue and possibly their own survival.

Furthermore, immune cells infiltrating the CNS have the potential to reorganize neuronal structures. Immune cell-derived mediators such as proinflammatory cytokines are capable of supporting neuronal regeneration, whereas impaired MHC class I-expressing neurons are potential targets of immune attack. Thus, immune cells and the immunological milieu can be both harmful and helpful to neurons. However, in most cases the direct contribution of the immune milieu to neuronal survival is still to be elucidated.

Summary

Microenvironmental factors have a profound influence on resident cell populations and their ability to modulate an immune response. The unique central nervous system (CNS) microenvironment has important effects in this regard, resulting in the establishment of immune privilege. In the CNS, neurons control the expression of "immunologically-relevant" molecules such as the major histocompatibility complex (MHC) and proinflammatory cytokines. This regulation of immune expression is predominantly mediated by inhibitory signals coming from electrically active neurons. For example, neurotrophins released by electrically active neurons have been identified as mediators involved in the downregulation of MHC class II on neighboring microglial cells, the main antigen-presenting cell type of the CNS.

Expression of immune effector molecules, such as proinflammatory cytokines, in response to neuronal damage could exert a variety of actions on the neurons themselves. In the first instance, activation of the local immune response could be harmful to resident CNS cells.

Alternatively, immune cell-derived mediators could protect and support the regeneration of damaged neurons. Thus, neurons communicate with the immune system by preventing immune activation in healthy CNS tissue, or permitting immune responses in processes that interfere with neuronal integrity. As a result, activation of the immune system by damaged neurons might help to reorganize the impaired CNS tissue.

Acknowledgement
We thank Ingeborg Haarmann and Dr. Alexander Flügel for help with the experiments involving GFP-transduced T lymphocytes and Prof. Wekerle for continuous support. Work in our laboratory was supported by DFG (SFB 391) and VW-Stiftung.

References

1. Streit WJ, Graeber MB, Kreutzberg GW (1989) Peripheral nerve lesion produces increased levels of major histocompatibility complex antigens in the central nervous system. J Neuroimmunol 21:117-123
2. Finsen BR, Tönder N, Xavier GF, Sörensen JC, Zimmer J (1993) Induction of microglial immunomolecules by anterogradely degenerating mossy fibers in the rat hippocampal formation. J Chem Neuroanat 6:276-275
3. Raivich G, Jones LL, Kloss CUA, Werner A, Neumann H, Kreutzberg GW (1998) Immune surveillance in the injured nervous system: T lymphocytes invade the axotomized mouse facial motor nucleus and aggregate around sites of neuronal degeneration. J Neurosci 18:5804-5816
4. Maehlen J, Olsson T, Zachau A, Klareskog L, Kristenssen K (1989) Local enhancement of major histocompatibility complex (MHC) class I and class II expression and cell infiltration in experimental allergic encephalomyelitis around axotomized motor neurons. J Neuroimmunol 23:125-132
5. Hickey WF (1991) Migration of hematogenous cells through the blood-brain barrier and the initiation of CNS inflammation. Brain Pathol 1:97-106
6. Tchélingérian J-L, Quinonero J, Booss J, Jacque C (1993) Localization of TNF-α and IL-1β immunoreactivities in striatal neurons after surgical injury to the hippocampus. Neuron 10:213-224
7. Neumann H, Schmidt H, Wilharm E, Behrens L, Wekerle H (1997) Interferon-γ gene expression in sensory neurons: Evidence for autocrine gene regulation. J Exp Med 186:2023-2031
8. Neumann H, Boucraut J, Hahnel C, Misgeld T, Wekerle H (1996) Neuronal control of MHC class II inducibility in rat astrocytes and microglia. Eur J Neurosci 8:2582-2590

9. Dani JW, Chernjavski A, Smith SJ (1992) Neuronal activity triggers calcium waves in hippocampal astrocyte networks. Neuron 8:429-440
10. Lee SC, Collins M, Vanguri P, Shin ML (1992) Glutamate differentially inhibits the expression of class II MHC antigens on astrocytes and microglia. J Immunol 148:3391-3397
11. Frohman EM, Vayuvegula B, Gupta S, Van den Noort S (1988) Norepinephrine inhibits γ-interferon-induced histocompatibility class II (Ia) antigen expression on cultured astrocytes via β2-adrenergic signal transduction mechanisms. Proc Natl Acad Sci USA 85:1292-1296
12. Frohman EM, Frohman TC, Vayuvegula B, Gupta S, Van den Noort S (1988) Vasoactive intestinal polypeptide inhibits the expression of the MHC class II antigens on astrocytes. J Neurol Sci 88:339-346
13. Neumann H, Misgeld T, Matsumuro K, Wekerle H (1998) Neurotrophins inhibit class II inducibility of microglia: Involvement of the p75 receptor. Proc Natl Acad Sci USA 95:5779-5784
14. Thoenen H (1995) Neurotrophins and neuronal plasticity. Science 270:593-598
15. Neumann H, Cavalié A, Jenne DE, Wekerle H (1995) Induction of MHC class I genes in neurons. Science 269:549-552
16. Neumann H, Schmidt H, Cavalié A, Jenne D, Wekerle H (1997) MHC class I gene expression in single neurons of the central nervous system: Differential regulation by interferon-γ and tumor necrosis factor-α. J Exp Med 185:305-316
17. Corriveau RA, Huh GS, Shatz CJ (1998) Regulation of class I MHC gene expression in the developing and mature CNS by neural activity. Neuron 21:505-520
18. Neumann H, Wekerle H (1998) Neuronal control of the immune response in the central nervous system: Linking brain immunity to neurodegeneration. J Neuropathol Exp Neurol 58:1-9
19. Albert ML, Darnell JC, Bender A, Francisco LM, Bhardwaj N, Darnell RB (1998) Tumor-specific killer cells in paraneoplastic cerebellar degeneration. Nat Med 4:1321-1324
20. Hickey WF, Ueno K, Hiserodt JC, Schmidt RE (1992) Exogenously-induced, natural killer cell-mediated neuronal killing: A novel pathogenetic mechanism. J Exp Med 176:811-817
21. Piani D, Spranger M, Frei K, Schaffner A, Fontana A (1992) Macrophage-induced cytotoxicity of N-methyl-D-aspartate receptor positive neurons involves excitatory amino acids rather than reactive oxygen intermediates and cytokines. Eur J Immunol 22:2429-2436
22. Frade JM, Barde YA (1998) Microglia-derived nerve growth factor causes cell death in the developing retina. Neuron 20:35-41
23. Cheng B, Christakos S, Mattson MP (1994) Tumor necrosis factors protect neurons against metabolic-excitotoxic insults and promote maintenance of calcium homeostasis. Neuron 12:139-153
24. Rensing-Ehl A, Malipiero U, Irmler M, Tschopp J, Constam D, Fontana A

(1996) Neurons induced to express major histocompatibility complex class I antigen are killed via the perforin and not the Fas (Apo-1/CD95) pathway. Eur J Immunol 26:2271-2274

25. Rabchevsky AG, Streit WJ (1997) Grafting of cultured microglial cells into the lesioned spinal cord of adult rats enhances neurite outgrowth. J Neurosci Res 47:34-48

26. Rapalino O, Lazarov-Spiegler O, Agranov E, Velan GJ, Yoles E, Fraidakis M, Solomon A, Gepstein R, Katz A, Belkin M, Hadani M, Schwartz M (1998) Implantation of stimulated homologous macrophages results in partial recovery of paraplegic rats. Nat Med 4:814-821

27. Moalem G, Leibowitz-Amit R, Yoles E, Mor F, Cohen IR, Schwartz M (1999) Autoimmune T cells protect neurons from secondary degeneration after central nervous system axotomy. Nat Med 5:49-55

28. Lindholm D, Heumann R, Meyer M, Thoenen H (1987) Interleukin-1 regulates synthesis of nerve growth factor in non-neuronal cells of rat sciatic nerve. Nature 330:658-659

29. Heese K, Hock C, Otten U (1998) Inflammatory signals induce neurotrophin expression in human microglia cells. J Neurochem 70:699-707

30. Bruce AJ, Boling W, Kindy MS, Peschon J, Kraemer PJ, Carpenter MK, Holtsberg FW, Mattson MP (1996) Altered neuronal and microglial responses to excitotoxic and ischemic brain injury in mice lacking TNF receptors. Nat Med 2:788-794

31. Schwartz M, Solomon A, Lavie V, Ben-Bassat S, Belkin M, Cohen A (1991) Tumor necrosis factor facilitates regeneration of injured central nervous system axons. Brain Res 545:334-338

32. Ehrhard PB, Erb P, Graumann U, Otten U (1993) Expression of nerve growth factor and nerve growth factor receptor tyrosine kinase Trk in activated CD4-positive T cell clones. Proc Natl Acad Sci USA 90:10984-10988

33. Elkabes S, DiCicco-Bloom EM, Black IB (1996) Brain microglia/macrophages express neurotrophins that selectively regulate microglial proliferation and function. J Neurosci 16:2508-2521

34. Kerschensteiner M, Gallmeier E, Behrens L, Klinkert WEF, Kolbeck R, Hoppe E, Stadelmann C, Lassmann H, Wekerle H, Hohlfeld R (1999) Activated human T cells, B cells and monocytes produce brain-derived neurotrophic factor (BDNF) in vitro and in brain lesions: A neuroprotective role of inflammation? J Exp Med 189:865-870

Levels of platelet-activating factor in cerebrospinal fluid and plasma of patients with relapsing-remitting multiple sclerosis

M. Arese, L. Callea, C. Ferrandi , F. Bussolino

Introduction

Platelet-activating factor (PAF), a mediator of homotypic and heterotypic cell-to-cell communication, activates inflammatory cells and lymphocytes through a seven-spanning transmembrane domain receptor [1, 2]. Following appropriate stimulation, it is produced and released by monocytes, neutrophils, endothelial cells and T lymphocytes [3-8]. It is also produced by neurons and glial cells stimulated by neurotransmitters and tumor necrosis factor (TNF)-α, respectively [9, 10]. Vascular endothelium is a key target for PAF. It modifies the barrier function of a monolayer of endothelial cells in vitro [11, 12], is a powerful vasopermeabilizing molecule in vivo [13, 14], and promotes leukocyte adhesion and transmigration [15-17]. High PAF concentrations are toxic for endothelial cells, causing vacuolization and marked formation of blebs [12, 18].

We have previously demonstrated that PAF levels were significantly higher in plasma and cerebrospinal fluid (CSF) of patients with relapsing-remitting (RR) multiple sclerosis (MS) studied by magnetic resonance imaging (MRI) when compared to healthy controls or patients with secondary progressive MS. We also found that both plasma and CSF levels correlated with the number of gadolinium-enhancing lesions, markers of blood-brain barrier (BBB) injury. In contrast peak levels did not correlate with the number of white matter lesions on MRI, nor with the expanded disability status score [19]. These data suggest that PAF is a mediator involved in the early phase of BBB damage but does not characterize the overall behavior of MS. Furthermore, these data explain the contradictory results in the literature regarding a role of PAF in MS and in experimental autoimmune encephalomyelitis [20-23].

In order to confirm that PAF appears concomitantly with the relapse of the disease, we have evaluated PAF levels in plasma and CSF and the number of gadolinium-enhancing lesions in two patients with RR MS at the onset of clinical exacerbation. The results show that this mediator peaks with the appearance of gadolinium leakage at MRI.

Materials and methods

Patients

Plasma and CSF PAF levels were studied in 2 female patients (aged 30 and 36 years with disease duration of 6 and 8 years, respectively) with RR MS diagnosed according to Poser et al. [24]. PAF levels were also measured in the plasma of 30 healthy subjects (14 females, 16 males; age range, 21-59 years) and in CSF of 6 neurologically healthy subjects (5 females, 1 male; age range, 32-54 years) who underwent lumbar puncture for epidural analgesia for orthopedic knee or leg surgery. Sample collection was done the day before MRI and at the moment of clinical exacerbation. Brain MRI before and after intravenous gadoteridol injection was performed as previously described [19].

Extraction, characterization and measurement of PAF

Plasma from heparinized blood (centrifuged 3 min at 15 000 g, 4 °C) and CSF were rapidly mixed with methanol containing 50 mM acetic acid (1:5 v/v) to destroy the specific acetylhydrolase that inactivates PAF [25]. After a 20-min incubation at room temperature, samples were centrifuged at 3 000 g for 20 min. To each milliliter of supernatant phase, 0.1 ml methanol, 0.2 ml water and 2 ml chloroform were added. After a 30-min incubation at room temperature, the lower organic phase was recovered and dried under nitrogen. Samples were separated by thin layer chromatography (solvent system: chloroform/methanol/water, 65:35:6 v/v) [7]. Lipids having the same retention front as synthetic PAF were extracted and further purified by high pressure liquid chromatography (Applied Biosystem) equipped with a reverse-phase column (Spherisorb C18, 5 μm, 100 mm length x 1 mm internal diameter). The sample was eluted with methanol/isopropanol/hexane/0.1 M aqueous ammonium acetate (100:10:2:5 v/v) at a flow rate of 75 μl/min. PAF was measured by a biological assay based on washed rabbit platelet aggregation, using a calibration curve with synthetic PAF for each assay series [7]. Specificity of platelet aggregation was inferred from the inhibitory effect of 3×10^{-6} M WEB2170 (Boehringer Ingelheim), a PAF receptor antagonist. The lower limit of PAF detection was 7.6×10^{-12} M. Each sample was tested in triplicate with 3 independent platelet preparations and the SEM was < 4%. The physicochemical characteristics of PAF and its sensitivity to lipase were studied as previously described [7].

Results and discussion

Two patients with clinically defined RR MS were followed for two years, from either the fourth or sixth year of disease. During this time interval, patients A and B had respectively three and two relapses with appearance of gadolinium-enhancing lesions visualized by MRI (Figs 1 and 2). During the relapses, the PAF levels of patient B in plasma and CSF were higher than that of healthy controls (plasma, 0.48±0.36 ng/ml, n=30; CSF, 0.01±0.04 ng/ml, n=6). Patient A shared a similar behavior, but in one relapse episode we failed to detect the PAF increase in CSF (Fig. 2). During the remitting phase, PAF was not increased in the plasma of either patient (Fig. 1).

Our previous observations indicated that the levels of PAF in the biological fluids of RR MS and chronic progressive MS patients correlated with the clinical phase of disease and, in RR MS, with gadolinium enhancement on MRI, a marker of BBB disturbance. However, this correlation was not evident when PAF levels were compared with the number of white matter lesions on MRI or with the estimated disability status scale [19]. The observation that PAF levels in plasma and in CSF peak with the appearance of gadolinium enhancement support our previous observations indicating that PAF was instrumental in BBB damage.

The early events of MS are largely unknown, although there is much evidence that perivascular inflammation followed by breakdown of the BBB precedes structural damage [26-29].

This early BBB disruption in MS could be due to a variety of mechanisms, although it is reasonable to suppose that soluble mediators (autacoids, cytokines, proteases) released by inflammatory cells are directly responsible for BBB leakage and injury. There is much experimental evidence in favor of this view. Extravasation of mononuclear cells and lymphocytes into the central nervous system requires involvement of specific ligands (very late antigen-4, lymphocyte function associated antigen-1, sialylated Lewis-X antigens) with their counter-receptors (vascular cell adhesion molecule-1, intercellular adhesion molecule-1, selectins) [30, 31]. Vascular cell adhesion molecule-1 is up-regulated in MS brain tissue [32], and high levels of the soluble form of intercellular adhesion molecule-1, vascular cell adhesion molecule-1 and E-selectin are found in the

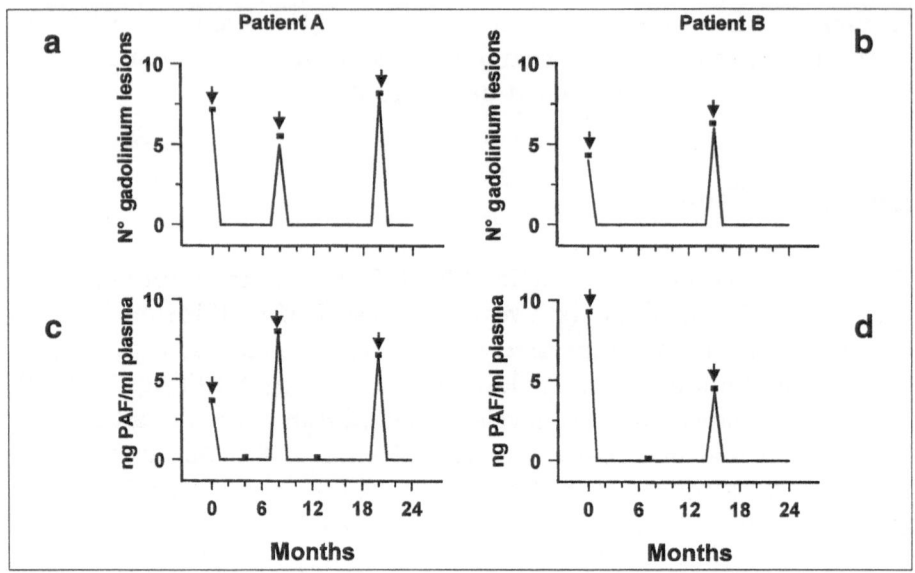

Fig. 1a-d. Number of gadolinium-enhancing lesions on MRI (a and b) and plasmatic PAF levels (c and d) in two patients affected by RR MS. *Arrows* indicate the clinical relapse. *Black squares* indicate plasma collection for PAF measurements

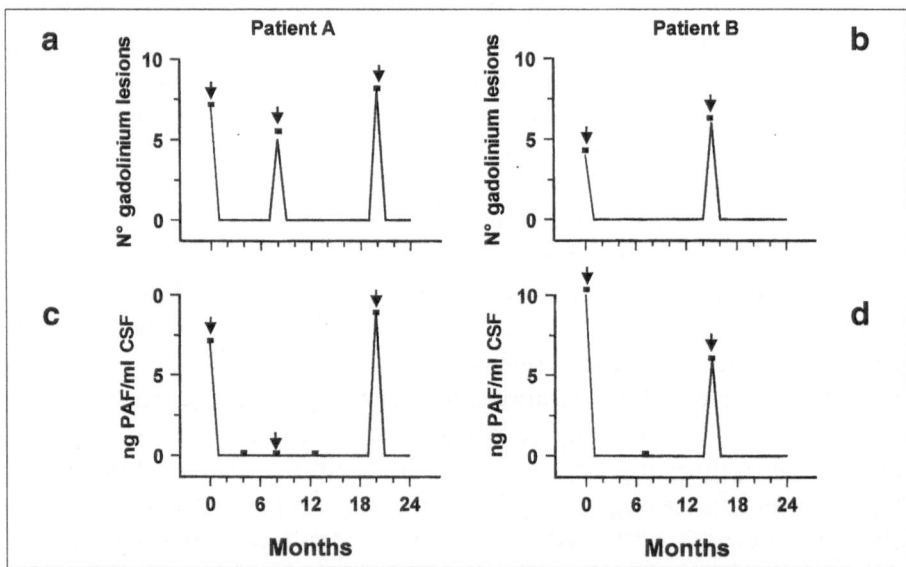

Fig. 2a-d. Number of gadolinium-enhancing lesions on MRI (a and b) and PAF levels in CSF (c and d) in two patients affected by RR MS. *Arrows* indicate the clinical relapse. *Black squares* indicate CSF collection for PAF measurements

serum of MS patients [30, 33]. The serum level of vascular cell adhesion molecule-1 is higher in patients with relapsing-remitting course with gadolinium-enhanced lesions on MRI as opposed to secondary progressive MS [33]. In addition, MS-derived microvascular endothelial cells express higher intercellular adhesion molecule-1 levels than do cells from normal brain, and are highly adhesive for leukocytes [34]. Treatment with a monoclonal antibody against very late antigen-4 reduces the leukocyte recruitment in experimental allergic encephalomyelitis and leads to both clinical and pathological improvements in vessels [35]. As far as the soluble mediators are concerned, $CD8^+$ T cells from MS patients have been shown to release chemokines that attract monocytes and promote the release of matrix metalloproteinase-9, which damages the basal membrane of blood [36]. NO, which may result in BBB perturbation [37], may be involved in the pathogenesis of experimental allergic encephalomyelitis [38, 39] and MS [40]. Lastly, complement activation via the classic pathway may be instrumental in BBB damage and in the generation of chemoattractant anaphylotoxins [41].

The ability of PAF to induce vasopermeabilization and leukocyte infiltration [1] fits well with our experimental data and suggests that this mediator plays a role in the early phases of MS. The specific and early role of PAF in MS pathogenesis could explain the conflictual results obtained with the use of a PAF receptor antagonist in the management of MS [20, 21]. Employment of PAF antagonists, in light of the MRI evidence, could thus be envisaged.

Summary

PAF is a phospholipid mediator of inflammation with a wide range of biological activities, including alteration of the barrier function of the endothelium. We have previously demonstrated that this molecule is elevated in plasma and CSF in patients with RR MS. In addition, PAF levels correlated with the number of gadolinium-enhancing lesions at MRI in these patients. Here, we report the follow-up in two patients affected by RR MS. Our results show that PAF peaked in both biological fluids, concomitantly to the increased number of gadolinium-enhancing lesions on MRI. These findings suggest that PAF is a significant mediator of blood-brain barrier injury which characterizes the onset of MS.

Acknowledgment
This work was supported by a grant from the Istituto Superiore della Sanità (Multiple Sclerosis Project) and from Fondazione Italiana per Sclerosi Multipla.

References

1. Bussolino F, Camussi G (1995) Platelet-activating factor produced by endothelial cells. A molecule with autocrine and paracrine properties. Eur J Biochem 229:327-337
2. McManus LM, Woodard DS, Deavers SI, Pinckard RN (1993) PAF molecular heterogeneity: pathobiological implications. Lab Invest 69:639-650
3. Lotner GZ, Lynch JM, Betz SJ, Henson PM (1980) Human neutrophil-derived platelet activating factor. J Immunol 124:676-684
4. Camussi G, Aglietta M, Coda R, Bussolino F, Piacibello W, Tetta C (1981) Release of platelet-activating factor (PAF) and histamine. II. The cellular origin of human PAF: monocytes, polymorphonuclear neutrophils and basophils. Immunology 42:191-199
5. Camussi G, Aglietta M, Malavasi F, Tetta C, Piacibello W, Sanavio F, Bussolino F (1983) The release of platelet activating factor from human endothelial cells in culture. J Immunol 131:2397-2403
6. Camussi G, Bussolino F, Salvidio G, Baglioni C (1987) Tumor necrosis factor/cachectin stimulates peritoneal macrophages, polymorphonuclear neutrophils, and vascular endothelial cells to synthesize and release platelet-activating factor. J Exp Med 166:1390-1404
7. Bussolino F, Camussi G, Baglioni C (1988) Synthesis and release of platelet-activating factor by human vascular endothelial cells treated with tumor necrosis factor or interleukin-1α. J Biol Chem 263:11856-11861
8. Le Gouvello S, Vivier E, Debre P, Thomas Y, Colard O (1992) CD2 triggering stimulates the formation of platelet activating factor-acether from alkyl-arachinoyl-glycerophosphocholine in a human CD4+ T lymphocyte clone. J Immunol 149:1289-1293
9. Jaranowska A, Bussolino F, Sogos V, Arese M, Lauro GM, Gremo F (1995) Platelet-activating factor production by human fetal microglia. Effect of lipopolysaccharides and tumor necrosis factor-a. Mol Chem Neuropathol 24:95-106
10. Sogos V, Bussolino F, Pilia E, Torelli S, Gremo F (1990) Acetylcholine induced production of platelet activating factor by human fetal brain cells in culture. J Neurosci Res 27:706-711
11. Bussolino F, Silvagno F, Garbarino G, Costamagna C, Sanavio F, Arese M, Soldi R, Aglietta M, Pescarmona G, Camussi G, Bosia A (1994) Human endothelial cells are targets for platelet-activating factor (PAF). Activation of alpha and beta protein kinase C isozymes in endothelial cells stimulated

by PAF. J Biol Chem 269:2877-2886

12. Bussolino F, Camussi G, Aglietta M, Braquet P, Bosia A, Pescarmona G, Sanavio F, D'Urso N, Marchisio PC (1987) Human endothelial cells are targets for endothelial cells. I. Platelet activating factor induces changes in cytoskeleton structure. J Immunol 139:2439-2446

13. Handley DA, Arbeeny CM, Lee ML, van Valen RG, Saunders RN (1984) Effect of platelet activating factor on endothelial permeability to macromolecules. Immunopharmacology 8:137-144

14. Humphrey DM, McManus LM, Satouchi K, Hanahan DJ, Pinckard RN (1982) Vasoactives properties of acetyl glyceryl ether phosphorylcholine and analogs. Lab Invest 46:422-427

15. Kuijpers TW, Hakkert BC, Hart MHL, Roos D (1992) Neutrophil migration across monolayers of cytokine-prestimulated endothelial cells: a role for platelet-activating factor and IL-8. J Cell Biol 117:565-572

16. Lorant DE, Topham MK, Whatley RE, McEver RP, McIntyre TM, Prescott SM, Zimmerman GA (1993) Inflammatory roles of P-selectin. J Clin Invest 92:559-570

17. Renkonen R, Mattila P, Ustinov J (1990) Signal transduction during platelet-activating factor-induced lymphocyte binding to endothelial cells. Scand J Immunol 31:523-527

18. Bourgain RH, Mael L, Braquet P, Andries R, Touqui L, Braquet M (1985) The effect of 1-O-alkyl-2-acetyl-sn-glycero-3-phosphocholine on the arterial wall. Prostaglandins 30:185-197

19. Callea L, Arese M, Orlandini A, Bargnani C, Priori A, Bussolino F (1999) Platelet activating factor is elevated in cerebral spinal fluid and plasma or patients with relapsing-remitting multiple sclerosis. J Immunol 94:212-221

20. Brochet B, Orgogozo JM, Guinot P, Dartigues JF, Henry P, Loiseau P (1992) Pilot study of ginkgolide B, a PAF-acether specific inhibitor in the treatment of acute outbreaks of multiple sclerosis. Rev Neurol 148:299-301

21. Brochet B, Guinot P, Orgogozo JM, Conavreux C, Rumbach L, Lavergne V (1995) Double blind placebo controlled multicentre study of ginkgolide B in treatment of acute exacerbations of multiple sclerosis. The Ginkgolide Study Group in Multiple Sclerosis. J Neurol Neurosurg Psychiatry 58:360-362

22. Howat DW, Chand N, Braquet P, Willoughby DA (1989) An investigation into the possible involvement of platelet activating factor in experimental allergic encephalomyelitis in rats. Agents Actions 27:473-476

23. Vela L, Garcia Merino A, Fernandez-Gallardo S, Sanchez Crespo M, Lopez Lozano JJ, Saus C (1991) Platelet-activating factor antagonists do not protect against the development of experimental autoimmune encephalomyelitis. J Neuroimmunol 33:81-86

24. Poser CM, Paty DW, Scheinberg L, McDonald WI, Davis FA, Ebers GC, Johnson KP, Sibley WA, Silberberg DH, Tourtellotte WW (1983) New diagnostic criteria for multiple sclerosis: guidelines for research protocols. Ann Neurol 13:227-231

25. Farr RS, Cox CO, Wardlow ML, Meng KE, Green DE (1983) Human serum acid-labile factor is an acetylhydrolase that inactivates platelet-activating factor. Fed Proc 42:3120-3122
26. Adams CWM, Poston RN, Burk SJ, Sidhu YS, Vipond H (1985) Inflammatory vasculitis in multiple sclerosis. J Neurol Sci 69:269-283
27. Guseo A, Jellinger K (1975) The significance of perivascular infiltrations in multiple sclerosis. J Neurol 211:51-60
28. Kermode AG, Thompson AJ, Tofts P, MacManus DG, Hendall BE, Kingsley DPE, Moseley IF, Rudge P, McDonald WI (1990) Breakdown of the blood-brain barrier precedes symptoms and other MRI signs of new lesions in multiple sclerosis. Brain 113:1477-1489
29. Moor ACE, DeVries HE, DeBoer AG, Breimer DD (1994) The blood-brain barrier and multiple sclerosis. Biochem Pharmacol 47:1717-1724
30. Dore-Duffy P, Newman W, Balabanov R, Lisak RP, Rothlein R, Peterson M (1995) Circulating, soluble adhesion proteins in cerebrospinal fluid and serum of patients with multiple sclerosis: correlation with clinical activity. Ann Neurol 37:55-62
31. Merril JE, Benveniste EN (1996) Cytokine in inflammatory brain lesions: helpful and harmful. Trends Neurosci 19:331-338
32. Verbeek MM, Westphal JR, Ruiter DJ, de Waal RM (1995) T lymphocyte adhesion to human brain pericytes is mediated via very late antigen-4/vascular cell adhesion molecule-1 interaction. J Immunol 154:5876-5884
33. Hartung HP, Reiners K, Archelos JJ, Michels M, Heidenreich F, Pflughaupt KW, Toyka KV (1995) Circulating adhesion molecules and tumor necrosis factor receptor in multiple sclerosis: correlation with magnetic resonance imaging. Ann Neurol 38:186-193
34. Lou J, Chofflon M, Juillard C, Donati Y, Mili N, Siegrist CA, Grau GE (1997) Brain microvascular endothelial cells and leukocytes derived from patients with multiple sclerosis exhibit increased adhesion capacity. Neuroreport 8:629-633
35. Soilu-Hanninen M, Roytta M, Salmi A, Salonen R (1997) Therapy with antibody against leukocyte integrin VLA-4 (CD49d) is effective and safe in virus facilitated experimental allergic encephalomyelitis. Neuroimmunology 72:95-105
36. Biddison WE, Cruikshank WW, Center DM, Connor EW, Honma K (1997) Chemokine and matrix metalloproteinase secretion by myelin proteolipid-specific CD8+ T cells: potential roles in inflammation. J Immunol 158:3046-3053
37. Hurst RD, Fritz IB (1996) Nitric oxide-induced perturbations in a cell culture model of the blood-brain barrier. J Cell Physiol 167:81-88
38. Lin RF, Lin TT, Tilton RG, Cross AH (1993) Nitric oxide localized to spinal cords of mice with experimental allergic encephalomyelitis: an electron paramagnetic study. J Exp Med 178:643-648
39. Cross AH, Misko TP, Lin RF, Hickey WF, Trotter JL, Tilton RG (1994) Aminoguanidine, an inhibitor of inducible nitric oxide synthase, amelio-

rates experimental autoimmune encephalomyelitis in SJL mice. J Clin Invest 93:2684-2690

40. Johnson AW, Land JM, Bolanos JP, Clark JB, Heales SJR (1995) Evidence for increased nitric oxide production in multiple sclerosis. J Neurol Neurosurg Psychiatry 58:107
41. Morgan BP, Gasque P (1996) Expression of complement in the brain: role in health and disease. Immunol Today 17:461-466

The contribution of magnetic resonance imaging to the understanding of multiple sclerosis pathogenesis

M. ROVARIS, G. IANNUCCI, M. FILIPPI

Introduction

Magnetic resonance imaging (MRI) is a sensitive tool for diagnosing and evaluating in vivo the dynamics of multiple sclerosis (MS) [1]. However, the low pathological specificity of conventional, T2-weighted MRI inevitably limits its potential for defining the pathophysiology of MS [1], although enhancement on T1-weighted scans after the injection of gadolinium-DTPA (Gd) can be used as a reliable marker of blood-brain barrier (BBB) dysfunction [1].

More recently, the introduction of new strategies for post-contrast imaging and the application of non-conventional techniques [e.g. magnetization transfer imaging (MTI) and magnetic resonance spectroscopy (MRS)] [2, 3] have improved our understanding of the evolution of MS lesions [1]. The present review will outline the major contributions which can be obtained by the use of MRI techniques in the study of MS pathogenesis.

Conventional MRI

Studies both in animals [4] and in human MS brain biopsy specimens [5, 6] demonstrated that Gd enhancement is consistent with histopathological findings of BBB breakdown. Perivascular inflammation appears to be a necessary precondition to the development of enhancement, since non-inflammatory demyelination is unaccompanied by changes in BBB permeability [7]. Longitudinal MRI studies confirm that enhancement occurs in almost all new lesions in patients with relapsing-remitting (RR) or secondary progressive (SP) MS [8]. Focal areas of increased signal on enhanced images can also be detected before the appearance of lesions on unenhanced T2-weighted scans [9]. Enhancement may reappear in older plaques, with or without a concomitant increase in their size [8], thus suggesting either partial repair of BBB or possible reactivation of BBB abnormalities.

The heterogeneity of BBB changes is reflected by the different morphological patterns of enhancement, i.e. nodular, patchy or ring-like. It has been suggested that nodular enhancing lesions represent small areas of perivascular inflammation either at the edges of established lesions or in areas of formerly normal appearing white matter (NAWM), whereas ring-enhancing areas are probably areas of acute inflammation at the edge of chronic demyelinated lesions [10]. Recently, Bruck et al. [11] and van Waesberghe et al. [12] found that ring enhancement is not restricted to reactivation of older MS lesions, but may be the first manifestation of new activity or the evolution of other enhancement patterns, especially in very large plaques.

Some insights about the various pathological substrates of enhancing lesions are also provided by the evolution of findings visible on unenhanced T1-weighted images. Most of the enhancing lesions show a corresponding area of hypointensity on pre-contrast T1-weighted images [12], whereas some lesions are isointense to the NAWM. On follow-up unenhanced T1-weighted scans, these lesions can be classified into four categories: (a) persistently isointense lesions, (b) temporarily isointense lesions, (c) persistently hypointense lesions, and (d) temporarily hypointense lesions [12]. Since the degree of hypointensity on T1-weighted scans is mainly related to the extent of extracellular oedema and axonal loss [13], pattern (a) may represent purely oedematous or inflammatory lesions, affected by only a minor degree of demyelination. Demyelination associated with mild loss of oligodendrocytes may be the histological phenotype of pattern (d), while persistently hypointense lesions (pattern c) are probably those in which axonal loss plays a relevant role in determining T1 hypointensity. At a 6-month follow-up [12], 75% of ring-enhancing lesions remain hypointense, thus suggesting that this pattern of enhancement may be a feature of lesions with a more severe involvement.

Recent strategies in post-contrast imaging have increased the sensitivity of MRI in the detection of enhancing lesions and have provided some insights about the pathological substrates of enhancing lesions. Using a triple dose (TD) of Gd, it has been shown that there are lesions which can be detected only after the administration of TD [14, 15], either because of the time course of their BBB leakage, implying that these lesions might be detectable only with TD for a part of the inflammatory episode, or because BBB permeability is too restricted for enhancement to be seen on standard dose (SD) scans, given the lower intravascular concentration of Gd. However, in a recent longitudinal MRI study [16], lesions enhancing only after TD had a shorter duration of their enhancement, while lesions enhancing after different Gd doses changed their pattern of enhancement on follow-up scans only in the minority of the cases (about 15%). In addition, in comparison to SD lesions, TD lesions had higher magnetization transfer ratio (MTR) values when they started to enhance, and higher degrees of MTR recovery during a 3-month follow-up period [17]. This suggests that the extent of BBB opening is correlated to the degree of associated tissue damage and that the magnitude of the two phenomena may vary in different enhancing MS lesions. These findings suggest that enhancing MS lesions form a heterogeneous population and those enhancing only after TD of Gd are characterized by a milder and shorter opening of the BBB.

Magnetization transfer imaging

MTI is a recent technique in which the image contrast depends on the interactions between protons in a relatively free state and those in a restricted motion state [2]. MTI studies of individual MS lesions [12, 17-19] confirm the pathological heterogeneity of T2-weighted abnormalities. Homogeneously enhancing lesions, which may represent new active lesions, have significantly higher MTR values than do ring-enhancing lesions [20], which may represent old, reactivated lesions. The duration of enhancement is also associated with different degrees of MTR changes in new MS lesions: lesions enhancing on at least two consecutive monthly scans have lower MTR than those enhancing on a single scan [21]. That a less damaged BBB is associated with milder tissue damage is also indicated by the demonstration that large enhancing lesions tend to have greater MTR reductions than smaller lesions [20].

Variable degrees of MTR changes in the NAWM may preceed the formation of enhancing lesions [22, 23]. These changes are detectable three months before lesion appearance and tend to become more evident on scans obtained closer to those where enhancement occurs.

Several longitudinal MTI studies have investigated the structural changes of new enhancing MS lesions [12, 17-23]. The results of all these studies consistently showed that, on average, MTR drops dramatically when the lesions start to enhance and may show a partial or complete recovery in the subsequent 1-6 months. However, only two studies [12, 18] evaluated the patterns of MTR evolution in individual lesions. In a study with monthly scans, vanWaesberghe et al. [12] showed that 44% of enhancing lesions had a marked MTR increase while 5% had a marked MTR decrease over a 6-month follow-up period, although the major changes occur in the first two months. The remaining 51% of the enhancing lesions studied had either a modest MTR increase or a modest MTR decrease. In a study of enhancing lesions from four patients followed for 9-12 months with monthly or trimonthly MTI scans, Dousset et al. [18] showed that 33% of lesions had a recovery of their MTR values which was close to the MTR of NAWM, 54% had an incomplete MTR recovery and 13% had a continuous worsening of their MTR. New lesions enhancing only after a TD injection of Gd had a similar short-term recovery profile [17]. However, at each time point of the follow-up, MTR in TD lesions was significantly higher than in SD lesions [17]. The most likely pathological mechanisms underlying these short-term MTR changes might be demyelination and remyelination. Such changes are highly variable from lesion to lesion and, therefore, different proportions of lesions with various degrees of structural changes might contribute to MS evolution. A recent longitudinal study [24] showed that newly enhancing lesions from patients with SP MS compared to those from patients with RR MS had lower MTR at the time of their appearance and presented a more severe and significant MTR reduction during the follow-up.

Recent studies suggest that measures obtained from MTI scans using whole-brain histogram analysis [25], which estimates the macro- and microscopic disease burden, are highly correlated with the extent of MS abnormalities on conventional scans [26] and with patients' clinical disability [27] and neuropsychological impairment [28]. Filippi et al. [27] reported that MTI histogram findings well differentiate MS clinical phenotypes, suggesting that, in patients with clinically isolated syndromes and benign MS, the pathology is mild and confined to the lesions visible on T2-weighted MRI. In the same study [27], it has also been shown that severe macro- and microscopic white matter damage is important for the development of MS disability, whereas a widespread, microscopic damage in the NAWM seems to be the major contributor to disability in patients with primary progressive MS.

Magnetic resonance spectroscopy

MRS allows in vivo identification of local changes in brain chemical composition, which may reflect the pathological evolution of MS lesions

[3]. Chronic MS lesions are characterized by reduced N-acetyl-aspartate/creatine (NAA/Cr) peaks [3], suggesting the presence of neuronal/axonal dysfunction, whereas acute lesions can present elevated choline (Cho) peaks, suggesting ongoing inflammation. MRS studies [29, 30] also showed that the amount of axonal loss within MS lesions may be important for the development of clinical disability. Davie et al. [29] found that, in patients with more disabling MS courses, NAA peaks from hyperintense lesions were significantly lower than those from lesions of patients with benign MS. Moreover, there was a significant, inverse correlation between the NAA concentration in MS lesions and the severity of clinical disability [29].

The application of MRS imaging (MRSI) [31-34] has recently enabled the study of brain pathology in visible MS lesions and in NAWM. Narayana et al. [33] found that spectroscopic changes in the NAWM, suggesting ongoing demyelination, may precede the appearance of visible MS lesions and may also occur in the absence of any discrete abnormality detected by conventional MRI. Fu et al. [34] found that differences in NAA/Cr ratios between RR MS and SP MS patients may result from the damage in the NAWM rather than from that due to macroscopic MS lesions. Changes in the NAWM NAA/Cr ratio correlated strongly with changes in clinical disability.

Conclusions

Conventional and non-conventional MRI techniques have significantly improved our understanding of MS pathogenesis. As regards the natural history of individual MS lesions, the following points have been established: (1) BBB dysfunction and inflammation are the most relevant aspects of the first stages of new lesion formation; (2) such changes may be preceeded by subtle changes in the NAWM; (3) the severity of the early changes significantly differs among different lesions; (4) the degree of recovery of white matter integrity after lesion formation is highly variable and ranges from lesion disappearance to irreversible tissue damage; and (5) in established lesions, progressive axonal loss and demyelination may occur and the relative proportions of "benign" and "malignant" lesions correlate with MS clinical evolution.

As regards MS evolution, the following conclusions can be drawn: (1) in the early phases of MS, the appearance of new lesions and their progressive accumulation over time are responsible for the acute disability and the uncomplete recovery after relapses; (2) the subsequent evolution, during the RR and SP phases of MS, involves not only lesion accumulation, but also further damage within visible lesions and NAWM; (3) if the latter processes are mild, the subsequent evolution is toward benign MS; and (4) microscopic tissue damage in the NAWM seems to be the major cause of clinical disability in patients with primary progressive MS.

References

1. Miller DH, Grossman RI, Reingold SC, McFarland HF (1998) The role of magnetic resonance techniques in understanding and managing multiple sclerosis. Brain 121:3-24
2. McGowan JC, Filippi M, Campi A, Grossman RI (1998) Magnetisation transfer imaging: theory and application to multiple sclerosis. J Neurol Neurosurg Psychiatry 64(Suppl 1):S66-S69
3. Arnold DL, Wolinsky JS, Matthews PM, Falini A (1998) The use of magnetic resonance spectroscopy in the evaluation of the natural history of multiple sclerosis. J Neurol Neurosurg Psychiatry 64(Suppl 1):S94-S101
4. Hawkins CP, Munro PMG, Mackenzie F et al (1990) Duration and selectivity of blood brain barrier breakdown in chronic relapsing experimental allergic encephalomyelitis studied by gadolinium-DTPA and protein markers. Brain 113:365-378
5. Katz D, Taubenberger JK, Raine C et al (1990) Gadolinium-enhancing lesions on magnetic resonance imaging: neuropathological findings. Ann Neurol 28:243
6. Nesbit GM, Forbes GS, Scheithauer BW et al (1991) Multiple sclerosis: histopathological and MR and/or CT correlation in 37 cases at biopsy and 3 cases at autopsy. Radiology 180:467-474
7. Dousset V, Brochet B, Vital A et al (1995) Lysolecithin-induced demyelination in primates: preliminary in vivo study with MR and magnetization transfer. AJNR Am J Neuroradiol 16:225-231
8. Miller DH, Rudge P, Johnson J et al (1988) Serial gadolinium-enhanced magnetic resonance imaging in multiple sclerosis. Brain 111:927-939
9. Kermode AG, Thompson AJ, Tofts PS et al (1990) Breakdown of the blood-brain barrier precedes symptoms and other MRI signs of new lesions in multiple sclerosis. Pathogenetic and clinical implications. Brain 113:1477-1489
10. Prineas JW, Connel F (1978) The fine structure of chronically active multiple sclerosis plaques. Neurology 28(Suppl):68-75
11. Bruck W, Bitsch A, Kolenda H, Bruck Y, Stiefel M, Lassmann H (1997) Inflammatory central nervous system demyelination: correlation of magnetic resonance imaging findings with lesion pathology. Ann Neurol 42:783-793

12. van Waesberghe JHTM, van Walderveen MAA, Castelijns JA et al (1998) Patterns of lesion development in multiple sclerosis: longitudinal observations with T1-weighted spin-echo and magnetisation transfer MR. AJNR Am J Neuroradiol 19:675-683

13. Barnes D, Munro PMG, Youl BD, et al (1991) The longstanding MS lesion. A quantitative MRI and electron microscopy study. Brain 114:1271-1280

14. Filippi M, Yousry T, Campi A et al (1996) Comparison of triple dose versus standard dose gadolinium-DTPA for detection of MRI enhancing lesions in patients with MS. Neurology 46:379-384

15. Filippi M, Rovaris M, Capra R et al (1998) A multi-centre longitudinal study comparing the sensitivity of monthly MRI after standard and triple dose gadolinium-DTPA for monitoring disease activity in multiple sclerosis: implications for clinical trials. Brain 121:2011-2020

16. Rovaris M, Mastronardo G, Gasperini C et al (1998) MRI evolution of new MS lesions enhancing after different doses of gadolinium. Acta Neurol Scand 98:90-93

17. Filippi M, Rocca MA, Rizzo G et al (1998) Magnetization transfer ratios in multiple sclerosis lesions enhancing after different doses of gadolinium. Neurology 50:1289-1293

18. Dousset V, Gayou A, Brochet B, Caille JM (1998) Early structural changes in acute MS lesions assessed by serial magnetization transfer studies. Neurology 51:1150-1155

19. Filippi M, Rocca MA, Horsfield MA, Comi G (1998) A one year study of new lesions in multiple sclerosis using monthly gadolinium-enhanced MRI. Correlations with changes of T2 and magnetization transfer lesion loads. J Neurol Sci 158:203-208

20. Silver NC, Lai M, Symms MR, Barker GJ, McDonald WI, Miller DH (1998) Serial magnetization transfer imaging to characterize the early evolution of new MS lesions. Neurology 51:758-764

21. Filippi M, Rocca MA, Comi G (1998) Magnetization transfer ratios of multiple sclerosis lesions with variable durations of enhancement. J Neurol Sci 159:162-165

22. Filippi M, Rocca MA, Martino G, Horsfield MA, Comi G (1998) Magnetization transfer changes in the normal appearing white matter precede the appearance of enhancing lesions in patients with multiple sclerosis. Ann Neurol 43:809-814

23. Goodkin DE, Rooney WD, Sloan R et al (1998) A serial study of new MS lesions and the white matter from which they arise. Neurology 51:1689-1697

24. Rocca MA, Mastronardo G, Rodegher M, Comi G, Filippi M (1999) Long term changes of MT-derived measures from patients with relapsing-remitting and secondary-progressive multiple sclerosis. AJNR Am J Neuroradiol (*in press*)

25. van Buchem MA, McGowan JC, Kolson DL, Polansky M, Grossman RI (1996) Quantitative volumetric magnetization transfer analysis in multiple

sclerosis: estimation of macroscopic and microscopic disease burden. Magn Reson Med 36:632-636

26. Phillips MD, Grossman RI, Miki Y et al (1998) Comparison of T2 lesion volume and magnetization transfer ratio histogram analysis and of atrophy and measures of lesion burden in patients with multiple sclerosis. AJNR Am J Neuroradiol 19:1055-1060

27. Filippi M, Iannucci G, Tortorella C et al (1999) Comparison of MS clinical phenotypes using conventional and magnetization transfer MRI. Neurology 52:588-594

28. Rovaris M, Filippi M, Falautano M et al (1998) Relation between MR abnormalities and patterns of cognitive impairment in multiple sclerosis. Neurology 50:1601-1608

29. Davie CA, Barker GJ, Thompson AJ, Tofts PS, McDonald WI, Miller DH (1997) H-1 magnetic resonance spectroscopy of chronic cerebral white matter lesions and normal appearing white matter in multiple sclerosis. J Neurol Neurosurg Psychiatry 63:736-742

30. Falini A, Calabrese G, Filippi M et al (1998) Benign versus secondary progressive multiple sclerosis: The potential role of proton MR spectroscopy in defining the nature of disability. AJNR Am J Neuroradiol 19:223-229

31. De Stefano N, Caramanos Z, Preul MC, Francis G, Antel JP, Arnold DL (1998) In vivo differentiation of astrocytic brain tumors and demyelinating lesions of the type seen in multiple sclerosis using H-1 magnetic resonance spectroscopic imaging. Ann Neurol 44:273-278

32. De Stefano N, Matthews PM, Fu LQ et al (1998) Axonal damage correlates with disability in patients with relapsing-remitting multiple sclerosis. Results of a longitudinal magnetic resonance spectroscopy study. Brain 121:1469-1477

33. Narayana PA, Doyle TJ, Lai DJ, Wolinsky JS (1998) Serial proton magnetic resonance spectroscopic imaging, contrast-enhanced magnetic resonance imaging and quantitative lesion volumetry in multiple sclerosis. Ann Neurol 43:56-71

34. Fu L, Matthews PM, De Stefano N et al (1998) Imaging axonal damage of normal-appearing white matter in multiple sclerosis. Brain 121:159-166

Chapter 8

The role of proinflammatory cytokines in multiple sclerosis

R. FURLAN, P.L. POLIANI, A. BERGAMI, M. GIRONI, G. DESINA, G. MARTINO

Introduction

The pathological hallmark of multiple sclerosis (MS) is the presence, within the central nervous system (CNS), of patchy inflammatory infiltrates leading to demyelination and axonal loss, and containing autoreactive T cells and pathogenic non-antigen-specific mononuclear cells [1]. It is currently believed that CNS antigen-reactive T cells provide the organ specificity of the pathogenic process. These cells regulate the recirculation within the CNS of non-antigen-specific lymphocytes and monocytes which act as effector cells by releasing myelinotoxic substances [2]. T cells specific for myelin and non-myelin components and mainly displaying the α/β T cell receptor (TCR) constitute the majority of the CNS-antigen specifc T cell population, while blood-borne activated macrophages, B cells producing antibodies against myelin components (i.e. myelin oligodendrocyte glycoprotein) or still unidentified components (i.e. oligoclonal cerebrospinal fluid bands), and γ/δ T cells represent the effector cell population. Nevertheless, the two different cell populations display overlapping functions; a minor proportion of α/β T cells specific for myelin antigens shows cytotoxic properties while γ/δ T cells can contribute to effector cell recruitment (mainly macrophages) via proinflammatory cytokine and chemokine production. To further complicate the T cell-mediated pathogenic scenario in MS, it has been recently reported that both regulatory as well as effector cells can be cross-regulated by different subsets of T cells including anti-T cell receptor (TCR) T cells as well as T cells carrying a natural killer receptor (NKR) [3].

MS is therefore considered an autoreactive T cell-mediated chronic inflammatory demyelinating disease of the CNS maintained by an immune-mediated reaction caused by a still unknown stimulus (i.e. virus). The whole process is potentially triggered by an inflammatory stimulus igniting a cascade of inflammatory events first occurring at the CNS level and then also involving systemic immune reactions which perpetuate the pathogenic mechanism [4]. Studies performed on transgenic animals affected by experimental autoimmune encephalomyelitis (EAE), the animal model for MS, have confirmed that not only CNS-specific T cells but also peripheral polyclonal expansion of lymphocytes driven by non-specific inflammatory stimuli are actually required to obtain CNS perivascular inflammatory infiltration and demyelination [5, 6]. However, the exact mechanism sustaining the pathogenic process in MS is still only partially defined. Among the putative mechanisms, possibly operating either at the CNS or peripheral level, those ignited and sustained by proinflammatory cytokines are considered essential. Proinflammatory cytokines can, in fact, (a) sustain the primary CNS-confined inflammatory process leading to the development of myelin-specific T cells, (b) activate myelin-specific T cells and shape their repertoire (Th1 vs. Th2 pattern), and (c) induce the recruitment of peripheral non-antigen-specific T cells as well as myelinotoxic effector cells, such as monocyte/macrophages, from the periphery to the CNS. Finally, certain proinflammatory cytokines determine direct oligodendrotoxicity (i.e. TNF-α) [7].

Cytokines and central nervous system inflammation

An immune-reaction occurring in the CNS [1-3] induces the release of primary inflammatory cytokines which represent the starting point of the pathological process occurring in MS. Tumor necrosis factor (TNF)-α, interleukin (IL)-1β, and IL-6 have been found in MS plaques at both the protein and mRNA levels [8]. When the so-called primary inflammatory cytokines are produced within the CNS, the inflammatory "MS-specific" process procedes via the in situ production of primary inflammatory cytokine-induced mediators. Among them, chemokines [8, 9], colony stimulating factors [10] and lipids [11-13] are the most important and have been found in active MS plaques. Due to the combined action of cytokines and chemokines, leukocytes are then recruited within the CNS. The typical pathological aspect of autoptic or brain biopsy material from MS patients indicates that lymphocytes and monocytes predominate in areas of demyelination and perivascular inflammation [14]. The recruitment of leukocytes within the CNS amplifies the local innate immunity via the activation of glial cells (mainly astrocytes and microglial cells), which then actively participate in the ongoing inflammatory process. Activated microglia form the main population of phagocytes in the early stage of demyelination [15, 16], and primary inflammatory cytokines are found in astrocytes at the edge of MS plaques [17]. In vivo evidence suggesting that a primary inflammatory, CNS-confined event may lead to perivascular mononuclear cell infiltration and demyelination comes from a recent study of MS patients in which we demonstrated that discrete areas of the normal-appearing white matter that subsequently will be infiltrated show magnetic resonance imaging changes *before* the appearance of the inflammatory foci [18].

Cytokines and CNS-antigen specific T cells

CNS-confined, innate immunity triggered by a still unknown primary inflammatory stimulus induces the development of T cells specific for myelin components. Primary inflammatory cytokines (i.e. IL-1/TNF) can induce, per se, myelin breakdown [7] the release of immunogenic myelin components and, in turn, a T cell-specific response, possibly occurring in the regional cervical lymph nodes [3]. T cells against myelin associated glycoprotein (MAG) [19], myelin basic protein (MBP) [20], proteolipid protein (PLP) [21] and myelin oligodendrocyte glyco-protein (MOG) [22] are present at higher levels compared to healthy subjects, both in the blood and cerebrospinal fluid (CSF) [23-25]. Beside macrophages, the main source of proinflammatory cytokines, antigen-specific T cells (mainly CD4+ but also CD8+) also produce a significant amount of these molecules.

CD4+ (and CD8+) T cells can be divided in three different subsets according to their cytokine secretion pattern. T helper 1 (Th1) cells secrete IL-2, interferon (IFN)-γ, and TNF-β, whereas T helper 2 (Th2) cells produce IL-4, IL-5, and IL-10. Their precursor cells, termed Th0 cells, can produce IL-2, IFN-γ and IL-4 simultaneously [26-28]. The cytokines elaborated by each Th-cell subset are inhibitory for the opposite subset. Th2 cytokines play a key role in immediate-type hypersensitivity; IL-4 is in fact the critical stimulus inducing a switch to IgE antibody production. Conversely, Th1 cytokines whose differentiation is mostly due to IL-12 stimulation, play a role in activation of macrophages (proinflammatory function) to kill intracellular parasites, in delayed hypersensitivity (DTH) and in the synthesis of opsonizing antibodies [29].

While the role of Th1 cytokines in the induction of experimental autoimmune diseases (i.e. EAE, insulin-dependent diabetes mellitus, col-lagen-induced arthritis [30-32]) has been widely documented, no clear-cut evidence that the same occurs in human organ-specific autoimmune diseases (i.e. Hashimoto's thyroiditis, Graves' disease, rheumatoid arthritis, and MS [33]) has been put forward. To further investigate the role of Th1/Th2 cells in MS, extensive analyses of the cytokine production pattern of MBP-reactive as well as PLP-reactive CD4$^+$ T cell lines (TCL) have been conducted. MBP-specific TCL can elaborate either an array of cytokines (IFN-γ and TNF-α) consistent with a Th1 pattern with no production of IL-4 and IL-5, or a set of Th2-cytokines (IL-4 and IL-5) with no or very little production of IFN-γ and TNF-α [34-37]. Different cytokine secretion patterns have also been observed among MBP-specific TCL generated from individuals subjects [34]. Further analyses of the cytokine production pattern of MBP-specifc T cell clones (TCC) yielded similar results: TCC were shown to produce both Th1- and Th2-like cytokines (Th0-like), high Th1 and low Th2 cytokines, or high Th2 and low Th1 cytokines. In contrast, PLP-specific TCC retrieved during acute attacks of MS, displayed Th1-like profiles, while TCC raised from the same patient during remission exhibited a Th0/Th2 pattern of cytokine secretion [38]. The heterogeneity of cytokine production patterns of myelin-specific TCC has been underscored by a recent study showing that alanine substitutions of critical TCR contact residues (90 and 91) of the MBP83-99 sequence change the cytokine profile pattern of these specific TCC [39]. It should however be noted that published results on the Th1/Th2 balance in MS may have been biased by the lack of patient stratification according to their disease course. Interestingly, it has recently been observed that a CD40 ligand-dependent Th1-type immune activation was found in progressive but not relapsing-remitting MS. A TCR-induced, CD40-mediated increase in IL-12 secretion was noted in progressive MS but not in relapsing-remitting MS patients in whom IL-12 release was similar to that measured in healthy controls [40]. Moreover, PLP-specific TCC generated from chronic progressive MS patients were able to proliferate and secrete cytokines in the absence of professional antigen presenters, and showed significantly enhanced IL-12-mediated proliferative response through a pathway independent of antigen processing [41].

In conclusion, although not supported by undisputed data, autoreactive T cells secreting proinflammatory cytokines (Th1) seem to exert a pathogenic action in MS during phases of disease activity either by improving the proinflammatory functions of these cells in an autocrine/paracrine fashion (i.e. macrophage activation) or by determining the recently described phenomenon called cytokine bystander T cell activation.

Systemic (e.g. bystander) effect of proinflammatory cytokines in MS

In MS patients, the local effects due to the inflammatory process are also paralleled by systemic effects which tend to amplify the local innate and specific immunities. Several non-MS specific inflammatory markers have been variably found in blood as well as in CSF samples from MS patients. Among them, primary inflammatory cytokines are consistently found and seem to correlate with preclinical phases of disease activity [42]. The presence of inflammatory cytokines in the peripheral circulation of patients with MS could contribute to amplify the CNS-confined inflammatory events. Myelin-specific T cells are in fact present in the peripheral circulation of normal individuals [43] as well as MS patients, and can be (re)activated via non-specific inflammatory stimuli such as primary inflammatory cytokines [44]. In non-human primates, MBP-specific T cells producing proinflammatory cytokines after in vitro activation are able to induce EAE upon transfer into irradiated autologous recipients [45]. MS patients experience clinical exacerbations when systemically treated with IFN-γ [46, 47]; circulating levels of TNF-α and IFN-γ increase during active phases of MS [48].

Myelin-reactive T cells are then pivotal in orchestrating and perpetuating the inflammatory events during MS. When the inflammatory process starts, it proceeds like a vicious circle. Myelin-specific T cells may produce Th1-type cytokines including IFN-γ which in turn increase the production and the activity of primary inflammatory cytokines either locally where these cytokines can directly destroy the myelin sheath (e.g. TNF-α) [7], or systemically where these cytokines can activate macrophages as well as T cells in a non-antigen-specific manner. This latter hypothesis is supported by the recent findings that (1) non-antigen-specific priming can be induced in vitro in CD4$^+$ T cells by proinflammatory cytokines (i.e. TNF-α, IL-2, and IL-6) [49] and in CD8$^+$ T cells by IFNs [50], and (2) that T cells (mainly CD4$^+$-memory) from MS patients can be activated in a myelin-independent fashion by the combination of two primary inflammatory cytokines (TNF-α and IL-6), the prototypical Th1 cytokine (i.e. IFN-γ and a T cell growth factor (i.e. IL-2). We have shown this cytokine-mediated "bystander" effect in MS patients and demonstrated that it is sustained by a cytokine-induced persistent increase of intracellular calcium. This effect is due to the stimulation of two different intracellular calcium-mediated pathways: one ignited by IFN-γ alone and mediated by protein kinase C [51], and the other ignited by the combination of IL-2, IL-6 and TNF-α and sustained by the increase of inositol-triphosphate production [44]. This latter pathway mediates the translocation of nuclear factor of activated T cells (NF-AT) into the T cell nucleus to initiate IL-2 gene transcription and, as a result, T cell activation [52]. This peripheral mechanism can facilitate the recrutiment of non-antigen-specific inflammatory cells (mainly lymphomonocytes), releasing myelinotoxic substances, to different CNS areas.

Conclusions

The role of proinflammatory cytokines in the pathogenic process underlying MS is exerted at different levels (Fig. 1). These molecules can orchestrate immune functions of either regulatory (myelin-specific T cells) or effector cells (monocyte/macrophages, B cells, non-antigen-specific

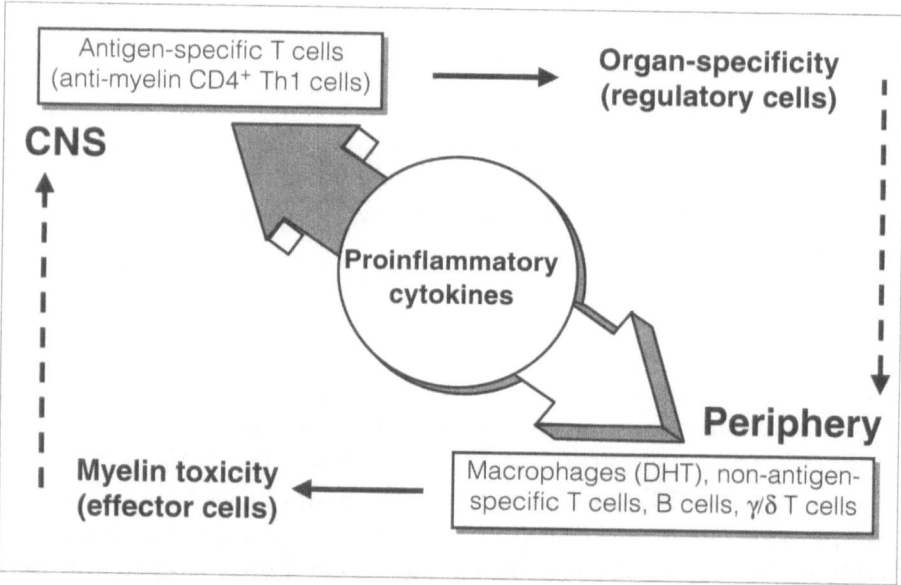

Fig. 1. Proinflammatory cytokines (i.e. IL-1α/β, IL-6, TNF-α/β, IFN-γ) play a central role in MS pathogenesis. They act on the two immune cell populations considered pivotal in the pathogenic MS scenario: (a) myelin-specific CNS-confined autoreactive T cells (possibly CD4+ Th1 cells), and (b) peripheral immune cells (mainly macrophages, B cells, non-antigen-specific T cells, γ/δ T cells) acting as effector cells by releasing myelinotoxic substances. Proinflammatory cytokines do not act only on blood-borne CNS-driven immune cells but also on CNS-resident cells, such as glial cells, thus representing a crucial player in the whole pathogenic mechanism sustaining and perpetuating MS. *DHT*, delayed-type hypersensivity

T cells) releasing myelinotoxic substances. Moreover, proinflammatory cytokines (mainly TNF-α) can exert a direct myelinotoxic effect or induce the secretion of myelinotoxic substances and of molecules able to increase T cell activation and CNS trafficking (i.e. metalloproteinases). Therefore, proinflammatory cytokines play a pivotal role in the pathogenic mechanisms sustaining MS, thus representing a suitable therapeutic target in this disease.

Acknowledgment

This work was supported in part by Istituto Superiore di Sanità (target project: Multiple Sclerosis) and MURST.

References

1. Martin R, McFarland HF, McFarlin DE (1992) Immunological aspects of demyelinating diseases. Annu Rev Immunol 10:153-187
2. Steinman L (1996) A few autoreactive cells in an autoimmune infiltrate control a vast population of nonspecific cells: a tale of smart bombs and the infantry. Proc Natl Acad Sci USA 93:2253-2256
3. Martino G, Hartung HP (1999) Immunopathogenesis of MS: the role of T cells. Curr Opin Neurol (in press)
4. Martino G, Furlan R, Poliani PL (1999) Inflammation in multiple sclerosis: a close interplay. In: Martino G, Adorini L (eds) From basic immunology to immune-mediated demyelination. Springer, Milan, pp 185-194 (Topics in neuroscience)
5. Goverman J, Woods A, Larson L, Weiner LP, Hood L, Zaller DM (1993) Transgenic mice that express a myelin basic protein-specific T cell receptor develop spontaneous autoimmunity. Cell 72:551-560
6. Lafaille JJ, Nagashima K, Katsuki M, Tonegawa S (1994) High incidence of spontaneous autoimmune encephalomyelitis in immunodeficient anti-myelin basic protein T cell receptor transgenic mice. Cell 78:399-408
7. Selmaj KW, CS Raine (1988) Tumor necrosis factor mediates myelin and oligodendrocyte damage in vitro. Ann Neurol 23:339-346
8. Esiri MM, Gay D (1997) The immunocytochemistry of multiple sclerosis plaques. In: Raine CS, McFarland HF, Tourtellotte WW (eds) Multiple sclerosis. Clinical and pathogenetic basis. Chapman & Hall, London, pp 173-186
9. Karpus WJ, Ransohoff RM (1998) Chemokine regulation of experimental autoimmune encephalomyelitis: temporal and spatial expression patterns govern disease pathogenesis. J Immunol 161:2667-2671

10. Battistini L, Borsellino G, Sawicki G, Poccia F, Salvetti M, Ristori G, Brosnan CF (1997) Phenotypic and cytokine analysis of human peripheral blood gamma delta T cells expressing NK cell receptors. J Immunol 159:3723-3730

11. Newcombe J, Li H, Cuzner ML (1994) Low density lipoprotein uptake by macrophages in multiple sclerosis plaques: implications for pathogenesis. Neuropathol Appl Neurobiol 20:152-162

12. Huterer SJ, Tourtellotte WW, Wherrett JR (1995) Alterations in the activity of phospholipases A2 in postmortem white matter from patients with multiple sclerosis. Neurochem Res 20:1335-1343

13. Narayana PA, Doyle TJ, Lai D, Wolinsky JS (1998) Serial proton magnetic resonance spectroscopic imaging, contrast-enhanced magnetic resonance imaging, and quantitative lesion volumetry in multiple sclerosis. Ann Neurol 43:56-71

14. Lucchinetti CF, Bruck W, Rodriguez M, Lassmann H (1996) Distinct patterns of multiple sclerosis pathology indicates heterogeneity on pathogenesis. Brain Pathol 6:259-274

15. Cuzner ML, Gveric D, Strand C, Loughlin AJ, Paemen L, Opdenakker G, Newcombe J (1996) The expression of tissue-type plasminogen activator, matrix metalloproteases and endogenous inhibitors in the central nervous system in multiple sclerosis: comparison of stages in lesion evolution. J Neuropathol Exp Neurol 55:1194-1204

16. Li H, Cuzner ML, Newcombe J (1996) Microglia-derived macrophages in early multiple sclerosis plaques. Neuropathol Appl Neurobiol 22:207-215

17. Selmaj KW, Raine CS, Cannella B, Brosnan CF (1991) Identification of lymphotoxin and tumor necrosis factor in multiple sclerosis lesions. J Clin Invest 87:949-954

18. Filippi M, Rocca MA, Martino G, Horshfield MA, Comi G (1998) Magnetization transfer changes in the normal-appearing white matter precede the appearance of enhancing lesions in patients with multiple sclerosis. Ann Neurol 43:809-814

19. Soderstrom M, Link H, Sun JB, Fredrikson S, Wang ZY, Huang WX (1994) Autoimmune T cell repertoire in optic neuritis and multiple sclerosis: T cells recognising multiple myelin proteins are accumulated in cerebrospinal fluid. J Neurol Neurosurg Psychiatry 57:544-551

20. Chou YK, Vainiene M, Whitham R, Bourdette D, Chou CH, Hashim G, Offner H, Vandenbark AA (1989) Response of human T lymphocyte lines to myelin basic protein: association of dominant epitopes with HLA class II restriction molecules. J Neurosci Res 23:207-216

21. Trotter JL, Hickey WF, van der Veen RC, Sulze L (1991) Peripheral blood mononuclear cells from multiple sclerosis patients recognize myelin proteolipid protein and selected peptides. J Neuroimmunol 33:55-62

22. Sun J, Link H, Olsson T, Xiao BG, Andersson G, Ekre HP, Linington C, Diener P (1991) T and B cell responses to myelin-oligodendrocyte glycoprotein in multiple sclerosis. J Immunol 146:1490-1495

23. Olsson T, Zhi WW, Hojeberg B, Kostulas V, Jiang YP, Anderson G, Ekre HP, Link H (1990) Autoreactive T lymphocytes in multiple sclerosis determined by antigen-induced secretion of interferon-gamma. J Clin Invest 86:981-985

24. Link H, Sun JB, Wang Z, Xu Z, Love A, Fredrikson S, Olsson T (1992) Virus-reactive and autoreactive T cells are accumulated in cerebrospinal fluid in multiple sclerosis. J Neuroimmunol 38:63-73

25. Sun JB, Olsson T, Wang WZ, Xiao BG, Kostulas V, Fredrikson S, Ekre HP, Link H (1991) Autoreactive T and B cells responding to myelin proteolipid protein in multiple sclerosis and controls. Eur J Immunol 21:1461-1468

26. Mosmann TR, Cherwinski H, Bond MW, Giedlin MA, Coffmann RL (1986) Two types of murine helper T cell clone. I. Definition according to profiles of lymphokine activities and secreted proteins. J Immunol 136:2348-2357

27. Del Prete G, De Carli M, Mastromauro C, Biagiotti R, Macchia D, Falagiani P, Ricci M, Romagnani S (1991) Purified protein derivative of *Mycobacterium tuberculosis* and excretory-secretory antigen(s) of *Toxocara canis* expand in vitro human T cells with stable and opposite (type 1 T helper or type 2 T helper) profile of cytokine production. J Clin Invest 88:346-350

28. Erard F, Wild M-T, Garcia-Sanz JA, Le Gros G (1993) Switch of CD8 T cells to noncytolytic CD8-CD4- cells that make TH2 cytokines and help B cells. Science 260:1802-1805

29. Romagnani S (1991) Human TH1 and TH2 subsets: doubt no more. Immunol Today 12:256-257

30. Baron JL, Madri JA, Ruddle NH, Hashim G, Janeway CA Jr (1993) Surface expression of alpha 4 integrin by CD4 T cells is required for their entry into brain parenchyma. J Exp Med 177:57-68

31. Haskins K, McDuffie M (1990) Acceleration of diabetes in young NOD mice with a CD4+ islet-specific T cell clone. Science 249:1433-1436

32. Nakajima H, Takamori H, Hiyama Y, Tsukada W (1990) The effect of treatment with interferon-gamma on type II collagen-induced arthritis. Clin Exp Immunol 81:441-445

33. Abbas AK, Murphy KM, Sher A (1996) Functional diversity of helper T lymphocytes. Nature 383:787-793

34. Hemmer B, Vergelli M, Tranquill L, Conlon P, Ling N, McFarland HF, Martin R (1997) Human T cell response to myelin basic protein peptide (83-99): extensive heterogeneity in antigen recognition, function and phenotype. Neurology 49:1116-1126

35. Voskuhl RR, Martin R, Bergman C, Dalal M, Ruddle NH, McFarland HF (1993) T helper 1 (TH1) functional phenotype of human myelin basic protein-specific T lymphocytes. Autoimmunity 15:137-143

36. Hemmer B, Vergelli M, Calabresi P, Huang T, McFarland HF, Martin R (1996) Cytokine phenotype of human autoreactive T cell clones specific for the immunodominant myelin basic protein peptide (83-99). J Neurosci Res 45:852-862

37. Hermans G, Stinissen P, Hauben L, Van den Berg-Loonen E, Raus J, Zhang J

(1997) Cytokine profile of myelin basic protein-reactive T cells in multiple sclerosis and healthy individuals. Ann Neurol 42:18-27

38. Correale J, Gilmore W, McMillan M, Li S, McCarthy K, Le T, Weiner LP (1995) Patterns of cytokine secretion by autoreactive proteolipid protein-specific T cell clones during the course of multiple sclerosis. J Immunol 154:2959-2968

39. Kozovska M, Zang YC, Aebischer I, Lnu S, Rivera VM, Crowe PD, Boehme SA, Zhang JZ (1998) T cell recognition motifs of an immunodominant peptide of myelin basic protein in patients with multiple sclerosis: structural requirements and clinical implications. Eur J Immunol 28:1894-1901

40. Balashov KE, Smith DR, Khoury SJ, Hafler DA, Weiner HL (1997) Increased interleukin-12 production in progressive multiple sclerosis: induction by activated CD4+ T cells via CD40 ligand. Proc Natl Acad Sci USA 94:599-603

41. Correale J, McMillan M, Li S, McCarthy K, Le T, Weiner LP (1997) Antigen presentation by autoreactive proteolipid protein peptide-specific T cell clones from chronic progressive multiple sclerosis patients: roles of costimulatory B7 molecules and IL-12. J Neuroimmunol 72:27-43

42. Martino G, Consiglio A, Franciotta DM, Corti A, Filippi M,Vandenbroeck K, Sciacca FL, Comi G, Grimaldi LME (1997) Tumor necrosis factor α and its receptors (R1 and R2) in relapsing-remitting multiple sclerosis. J Neurol Sci 152:51-61

43. Burns J, Rosenzweig A, Zweiman B, Lisak RP (1983) Isolation of myelin basic protein-reactive T cell lines from normal human blood. Cell Immunol 81:435-440

44. Martino G, Grohovaz F, Brambilla E, Codazzi F, Consiglio A, Clementi E, Filippi M, Comi G, Grimaldi LME (1998) Proinflammatory cytokines regulate antigen-independent T cell activation by two separate calcium-signaling pathways in multiple sclerosis patients. Ann Neurol 43:340-349

45. Meinl E, Hoch RM, Dornmair K, de Waal Malefyt R, Bontrop RE, Jonker M, Lassmann H, Hohlfeld R, Wekerle H, 't Hart BA (1997) Encephalitogenic potential of myelin basic protein-specific T cells isolated from normal rhesus macaques. Am J Pathol 150:445-453

46. Panitch HS, Hirsch RL, Haley AS, Johnson KP (1987) Exacerbations of multiple sclerosis in patients treated with gamma interferon. Lancet I:893-895

47. Panitch HS, Hirsch RL, Schindler J, Johnson KP (1987) Treatment of MS with gamma interferon: exacerbations associated with activation of the immune system. Neurology 37:1097-1102

48. Beck J, Rondot P, Catinot L, Falcoff E, Kirchner H, Wietzerbin J (1988) Increased production of interferon gamma and tumor necrosis factor precedes clinical manifestation in multiple sclerosis: do cytokines trigger off exacerbations? Acta Neurol Scand 78:318-323

49. Unutmaz D, Pileri P, Abrignani S (1994) Antigen-independent activation of naive and memory resting T cells by a cytokine combination. J Exp Med 180:1159-1164

50. Tough DF, Borrow P, Sprent J (1996) Induction of bystander T cell prolifer-

ation by viruses and type I interferon in vivo. Science 272:1947-1950

51. Martino G, Clementi E, Brambilla E, Moiola L, Comi G, Meldolesi J, Grimaldi LME (1994) γ-Interferon activates a previously undescribed Ca^{2+} influx in T lymphocytes from patients with multiple sclerosis. Proc Natl Acad Sci USA 91:4825-4829

52. Dolmetsch RE, Xu K, Lewis RS (1998) Calcium oscillations increase the efficiency and specificity of gene expression. Nature 392:933-936

Experimental autoimmune encephalomyelitis in the common marmoset *Callithrix jacchus*

A. Uccelli, G.L. Mancardi, D. Giunti, H. Brok, L. Roccatagliata, E. Capello, B. t'Hart

Introduction

Experimental autoimmune encephalomyelitis (EAE) has been utilized since the first studies of Rivers et al. [1] as the animal model of choice for autoimmune diseases and in particular demyelinating disease of the central nervous system (CNS) such as multiple sclerosis (MS). EAE is characterized by an autoimmune response against myelin antigens mediated mostly by T lymphocytes but also by macrophages and B cells [2]. Activated CD4+ T cells mediate EAE upon recognition of the target antigen presented by class II molecules of the major histocompatibility complex (MHC) [3]. EAE can be induced in susceptible species by means of active immunization with CNS proteins and by passive transfer of encephalitogenic T cells to syngeneic recipients. Encephalitogenic T cells recognize myelin antigens and can be retrieved from the blood of both immunized and naive animals [4].

Several species and strains have been utilized including rodents, rabbits and non-human primates. The clinical, pathological and immunological picture of EAE depends upon the mode of sensitization (primarily intradermally), the nature of the immunogen and the genetic background of the species utilized.

Immunogens include whole myelin homogenate as well as distinct myelin proteins, such as myelin basic protein (MBP), myelin oligodendrocyte glycoprotein (MOG) and proteolipid protein (PLP), emulsified with an equal volume of complete Freund's adjuvant containing *Mycobacterium tuberculosis* or *M. butyricum* to create an antigen depot. Boosts with *Bordetella pertussis* may be used to help in opening the blood-brain barrier (BBB). Depending upon the species, the antigen and the mode of sensitization, EAE has a wide array of clinical pictures that mimic human MS, such as a monophasic acute course, as well as a chronic relapsing course and even a primarily progressive course.

Pathology of EAE is characterized by perivascular infiltrates of inflammatory cells, mainly mononuclear cells, within the cerebral white matter. Depending on the immunizing antigen, it is possible to obtain a wide spectrum of neuropathological patterns inclusive of demyelination, remyelination, gliosis, loss of axons and, in certain species also necrosis [2].

Experimental autoimmune encephalomyelitis in marmosets

Although most studies on EAE have been performed in inbred species, the recent advances in housing and handling techniques, the increased knowledge of primate anatomy, immunology and genetics, and the compatibility of most of human reagents and diagnostic techniques have sparked wide interest in non-human primates. A unique advantage of monkeys arises from their outbred condition that closely resembles the human status. Irrespective of their outbred status, the transfer of immunocompetent cells in primates is allowed by the possibility of crossing the trans-species barrier among some closely related species [5] and by the natural bone marrow chimerism in others [4]. Therefore, as in inbred rodent strains, it is possible to elucidate the role of pathogenic cells by means of passive transfer experiments in a polymorphic setting similar to the human condition.

EAE has been induced in the common marmoset *Callithrix jacchus*, a small rat-sized neo-tropical primate species weighing 300-400 grams at adult age. Marmosets breed easily in captivity, giving birth to 1-2 non-identical sets of twin or triplet siblings per year [6]. Because siblings shared the placental blood circulation, bone marrow-derived cells were educated in a common thymic environment, making each twin tolerant to its fraternal Sibling's cells. This condition allows passive transfer of cells among genetically different siblings without eliciting an alloresponse [4]. Further similarity to the human condition is given by the close homology of marmoset's genes of the immune system. We recently demonstrated that *C. jacchus* T cell receptor (TCR) genes are evolutionary conserved [7] and that MHC class II region genes, despite a limited polymorphism, encode the evolutionary equivalents of the HLA-DR and -DQ molecules [8].

EAE has been induced in marmosets by active immunization with human white matter [6], with MOG, or with MBP followed by administration of MOG–specific antibodies [9]. Perivascular mononuclear cell infiltration, demyelination and reactive astrogliosis characterize EAE induced with whole myelin homogenate. As in human MS, early signs of axonal suffering can be detected (G.L. Mancardi, manuscript in preparation). Although administration of *B. pertussis* was utilized in the immunization protocol by Massacesi and coworkers [6], we observed that *Bordetella* may adversely affect the lesion pathology (B. 't Hart et al., unpublished observation). Using pre- and post-mortem magnetic resonance imaging (MRI), we studied the topography of EAE lesions in the marmoset brain and showed that it highly resembles that found in MS [10]. Lesions first occur mainly around the ventricles, but later also in the parenchyma of the white matter often by confluence of smaller perivascular lesions. Typical MS-like pathology can be achieved also by immunization with MOG but not by immunization with MBP or PLP. In the latter situation, mild inflammation without demyelination and little or no clinical signs were found [9]. Nevertheless, MBP-reactive T cells may play a role in EAE pathogenesis as demonstrated by the presence of proliferative responses to this antigen in animals immunized with whole myelin and by passive transfer experiments in which MBP-specific T cells induced a mild form of EAE [6].

Encephalitogenic T cells have been isolated from the blood of both naive and immunized animals, suggesting that autoreactive and potentially harmful lymphocytes are part of the normal marmoset immune system [4]. MBP-reactive T cells recognize different determinants by means of a diverse TCR repertoire. However, due to chronic antigen stimulation, clonal expansion of a few T cells may occur. Overall, the heterogeneity of the response to MBP is similar to that in humans and seems to be typical of outbred species as compared to inbred rodent strains where a limited response may be observed (A. Uccelli, manuscript in preparation).

The role of MOG is pivotal in the mechanisms leading to demyelination as was demonstrated by the necessity to administer anti-MOG antibodies after immunization with encephalitogenic T cells to obtain the complete spectrum of MS-like pathology [9]. These antibodies bind specifically to the disintegrating myelin membrane, suggesting their role in myelin degradation and phagocytosis [11]. In contrast to the broad response to MBP, T cells reacting against MOG seem to recognize a discrete number of epitopes. This may be related to the limited number of MHC restriction elements involved in the response to this antigen (H. Brok, manuscript in preparation). As in rodent EAE, intermolecular epitope spreading also occurs in marmosets. Following immunization with a chimeric dimer composed of MBP and PLP, a late antibody response against MOG appeared that correlated with the occurrence of demyelination [12]. Similarly, proliferative T cell responses to MOG but without antibody production have been detected in MBP-immunized marmosets (H. Brok, manuscript in preparation). The close association between MOG and demyelination has been emphasized by the detection of myelin degradation products within the urine of marmosets immunized with MOG. Such degradation products can be also detected in the urine of MS patients (J. Vogels et al., personal communication).

Activated T cells and macrophages appear to play important roles in situ, as shown by the high expression of CD40 and its ligand (CD154) in active lesions in marmosets [13]. Moreover, both Th1 and Th2 cytokines were detected in the same lesions, suggesting that plaques may arise from a complex interaction between proinflammatory and anti-inflammatory cytokines. The harmful role of Th1 cytokines in EAE pathogenesis has been demonstrated by the prevention of disease following treatment with the cAMP-specific type IV phosphodiesterase inhibitor rolipram which inhibits the release of tumor necrosis factor (TNF)-α by macrophages and monocytes [14]. Nevertheless, an exclusive anti-inflammatory role for Th2 cytokines is probably too simplistic as it arises from the worsening of EAE consequent to enhanced proliferative and antibody responses to MOG following immune deviation therapy [15].

Overall, EAE in *C. jacchus* is an useful model for studying the immunopathogenesis of autoimmunity within the CNS in an experimental setting that closely resembles the human condition. Moreover, the molecular and functional organization of the primate immune system, together with the possibility to utilize sophisticated diagnostic tools, leads to the possibility of evaluating the safety and efficacy of biological molecules as therapies for MS.

Acknowledgement
Some of the studies described in this chapter were funded by the "large scale facility" program from the European Community, by the Italian Society for Multiple Sclerosis (AISM) and by Istituto Superiore di Sanità, "Progetto Sclerosi Multipla".

References

1. Rivers TM, Sprunt DH, Berry GP (1933) Observations on the attempts to produce acute-disseminated encephalomyelitis in monkeys. J Exp Med

58:39-53

2. Lassmann H, Wekerle H (1998) Experimental models of multiple sclerosis. In: Compston A, Ebers G, Lassmann H, McDonald I, Matthews B, Wekerle H (eds) McAlpine's multiple sclerosis. Churchill Livingstone, London Edinburgh New York, pp 409-433

3. Zamvil SS, Steinman L (1990) The T lymphocyte in experimental allergic encephalomyelitis. Annu Rev Immunol 8:579-621

4. Genain CP, Lee-Parritz D, Nguyen MH, Massacesi L, Joshi N, Ferrante R, Hoffman K, Moseley M, Letvin NL, Hauser SL (1994) In healthy primates, circulating autoreactive T cells mediate autoimmune disease. J Clin Invest 94:1339-1345

5. Bontrop RE, Otting N, Slierendregt BL, Lanchbury JS (1995) Evolution of major histocompatibility complex polymorphisms and T cell receptor diversity in primates. Immunol Rev 143:43-62

6. Massacesi L, Genain CP, Lee-Parritz D Letvin NL, Canfield D, Hauser SL (1995) Chronic relapsing experimental autoimmune encephalomyelitis in new world primates. Ann Neurol 37:519-530

7. Uccelli A, Oksenberg JR, Jeong M, Genain CP, Rombos T, Jaeger E, Lanchbury J, Hauser SL (1997) Characterization of the TCRB chain repertoire in the New World monkey *Callithrix jacchus*. J Immunol 158:1201-1207

8. Antunes SG, de Groot NG, Brok H, Doxiadis G, Menezes AA, Otting N, Bontrop RE (1998) The common marmoset: a new world primate species with limited MHC class II variability. Proc Natl Acad Sci U S A 95:11745-11750

9. Genain CP, Nguyen MH, Letvin NL, Pearl R, Davis RL, Adelman L, Lees MB, Linington C, Hauser SL (1995) Antibody facilitation of multiple sclerosis-like lesions in a non-human primate. J Clin Invest 96:2966-2974

10. 't Hart BA, Bauer J, Muller HJ, Melchers B, Nicolay K, Brok H, Bontrop RE, Lassmann H, Massacesi L (1998) Histopathological characterization of magnetic resonance imaging-detectable brain white matter lesions in a primate model of multiple sclerosis: a correlative study in the experimental autoimmune encephalomyelitis model in common marmosets (*Callithrix jacchus*). Am J Pathol 153:649-663

11. Genain CP, Cannella B, Hauser SL, Raine CS (1999) Identification of autoantibodies associated with myelin damage in multiple sclerosis. Nat Med 5:170-175

12. McFarland HI, Lobito AA, Johnson MM, Nyswaner JT, Frank JA, Palardy GR, Tresser N, Genain CP, Mueller JP, Matis LA, Lenardo MJ (1999) Determinant spreading associated with demyelination in a nonhuman primate model of multiple sclerosis. J Immunol 162:2384-2390

13. Laman JD, van Meurs M, Schellekens MM, de Boer M, Melchers B, Massacesi L, Lassmann H, Claassen E, 't Hart BA (1998) Expression of accessory molecules and cytokines in acute EAE in marmoset monkeys (*Callithrix jacchus*). J Neuroimmunol 86:30-45

14. Genain CP, Roberts T, Davis R, Nguyen M, Uccelli A, Faulds D, Hoffman K, Timmel G, Li Y, Ferrante R, Joshi N, Hedgpeth J, Hauser SL (1995) Prevention of autoimmune demyelination in non-human primates by a cAMP-specific phosphodiesterase inhibitor. Proc Natl Acad Sci USA 92:3601-3605

15. Genain CP, Abel K, Belmar N, Villinger F, Rosenberg DP, Linington C, Raine CS, Hauser SL (1996) Late complications of immune deviation therapy in a nonhuman primate. Science 274:2054-2057

Subject Index